Cotton **Facts**

Cotton **Facts**

Technical Paper No. 25

of the

Common Fund for Commodities

M. Rafiq Chaudhry

Andrei Guitchounts

International Cotton Advisory Committee

Cotton Producing Areas in the World

Europe

Americas

Asia

Oceania

rica

by Carmen S. León

First Edition 2003   ISBN 0-9704918-3-2

Library of Congress Control No. 2003107604

Contents

Foreword

Cotton is a major agricultural and industrial crop that provides employment and income for hundreds of millions of people involved in its production, processing and marketing in more than 60 countries. For many countries in the developing world, cotton forms the economic base and is a major foreign exchange earner.

The Common Fund for Commodities and the International Cotton Advisory Committee, within their specific mandates, have established a fruitful co-operation in providing support to the cotton growers in developing countries through projects ranging from crop protection, cotton variety development and post-harvest activities to by-product development and market oriented activities.

Cotton Facts, which is an example of this co-operation, aims at providing a concise overview of definitions and terminology commonly used in cotton production, processing, marketing and use. The book contains a wealth of factual information and is directed to practitioners, researchers, scholars and students who wish to have a quick reference guide on specific cotton-related issues.

This publication has been prepared by senior staff of the International Cotton Advisory Committee, supported by information from their extensive network of technical and marketing experts.

It is hoped that this publication will be of service to all those who have an interest in production, consumption and trade of cotton.

Rolf W. Boehnke
Managing Director
Common Fund for Commodities

Terry P. Townsend
Executive Director
International Cotton Advisory Committee

Introduction

Cotton, unique among agricultural crops, provides food and fiber. Cotton is a major natural fiber crop, and also provides an edible oil and seed by-products for livestock food. However, most of the research on cotton and the treatment it receives in the world are limited to its recognition as a fiber crop. Cottonseed accounts for two-thirds of the weight of harvest, but all-out efforts are made to improve primarily the fiber produced.

Cotton has an impact on our lives beyond recognition, and genetic engineering technology has the potential to improve cotton further. Genetic engineering may even convert the cotton plant into a biofactory. Natural cotton fiber is a great gift from nature. Chemical fiber factories may close, but the cotton plant will continue to produce fiber and food for humans. Just imagine for a while if cotton did not exist in nature.

A lot of work has been done with great success to improve cotton productivity and fiber quality. But there are relatively few books published about cotton and much work remains un-published. The purpose of this publication, which was initiated by the International Cotton Advisory Committee and sponsored by the Common Fund for Commodities, is not to report new research on cotton which is done by researchers in various technical journals. The purpose of this publication is to collect, compile and publish hard facts on cotton based on work done by all cotton researchers, current and past. A collection of facts and definitions used in various stages of cotton production and marketing in one publication will serve as a useful reference tool for thousands of companies and individuals associated with the cotton industry.

Authors have tried to review as much literature as possible and collect facts about the cotton plant and its organs, agronomic and physiological behavior, breeding, pests, fiber, cotton trading, shipping, pricing and price risk management, international cotton organizations and associations and cotton exchanges and their history. But the greatest challenge for authors has been to decide what to include and what not to include as a fact related to cotton. We recognize that more could be added to this book, but size is one primary limitation. Further, given that the cotton plant is highly responsive to environmental conditions, some facts could change or could prove incorrect under different conditions.

Original researchers deserve credit for this publication. Their work resulted in the development of these "facts." This is a book which is for everybody in cotton.

The Authors

Acknowledgements

The title of the book, Cotton Facts, clearly reflects the contents of the book. The purpose of this book was not to invent new facts but rather to review and compile available facts about cotton. For this purpose, we have reviewed a lot of published literature. The list is too long and it is simply impossible to include all references in this book, and only a part of the literature has been quoted. We are greatly indebted to all who did the original work and published in various forms. This book is also based on the cumulative experience of over 40 years by both authors working exclusively on cotton. We are also thankful to our colleagues with whom we have worked in various capacities since the late 1970s.

The Authors

The Cotton Plant & its Organs

THE PLANT

The word "cotton" is derived from the Arabic word "*al qatan.*" No one knows how old the cotton plant is. One of the first archeological discoveries of cotton usage in the world is located in Pakistan at Mohenjo Daro. This site is over 5,000 years old. Present day breeding and selection are a continuation of the domestication process.

The cotton plant is a perennial (lasting many years) tree but has been domesticated to be grown as a pseudo-annual shrub. As a perennial tree, the cotton plant is able to grow year after year, producing flowers and bolls, as long as conditions are suitable for growth. The cotton plant is indeterminate in nature. It will not stop producing flowers and bolls as long as growing conditions are suitable.

The cotton plant has hairs (trichomes) and irregularly dotted pigment glands.

Wild species of cotton generally occur in frost-free areas of the subtropical and tropical regions. Freezing temperatures kill the protoplast of all cultivated species and most wild species. Therefore, cotton is a warm climate crop. It is planted in spring or early summer and harvested in late fall or early winter.

Cotton is a sun-loving plant but not a water-loving plant. Water requirements of cotton depend on weather conditions, but a successful cotton harvest requires at least 75 cm of rain or irrigation water on average. Although cotton is planted in both hemispheres, most of it is cultivated in the Northern Hemisphere. The time of planting in the Northern Hemisphere is the time of harvesting in the Southern Hemisphere. Cotton is primarily grown between 37°N and 32°S; however, its cultivation has been extended to 45°N in China.

Taxonomy

Cotton is a member of the order *Malvales*, family *Malvaceae*. This makes it a relative of such familiar garden plants as the mallow and Rose of Sharon.

COTTON'S FAMILY TREE	
Rank	**Main line of descent**
Order	Malvales
Family	Malvaceae
Tribe	Gossypieae
Genus	Gossypium

Gossypium plants are small trees, shrubs or sub-shrubs that grow in tropical and subtropical regions of Africa, Asia, Australia and America. The genus *Gossypium* consists of 50 wild and cultivated species. Forty-five of the species are diploid, having a 2n chromosome number equal to 26.

At least eight diploid genomes, designated A, B, C, D, E, F, G and K, are found in the genus *Gossypium*. The A genome is restricted to two species, *Gossypium arboreum*, and *G. herbaceum* of the Old World. The D genome consists of 14 currently recognized species of the New World, such as *G. thurberi* and *G. raimondii*. Interspecific (among species) hybridization within diploids and tetraploids has not produced any useful varieties.

Only four species of *Gossypium* are grown on a commercial scale in the world. Also known as cultivated species, they are *G. hirsutum* L., *G. barbadense* L., which are called New World species; and *G. arboreum* L. and *G. herbaceum* L., which are called Old World or Asiatic cottons. *G. arboreum* originated in the Indo-Pak sub continent. *G. herbaceum* originated in southern Africa. *G. barbadense* originated in Peru. *G. hirsutum* originated in Mexico. The origin of these four species, cultivated in four different areas of the world so far apart, indicates that they were domesticated independently of each other. *G. barbadense* is the most photoperiodic (sensitive to day length) species among the four cultivated species, which limits its cultivation to only a few countries.

The most commonly cultivated species of cotton in the world is *G. hirsutum* L. This and *G. barbadense* are the most important agricultural cottons. Both are allotetraploids (AADD genomes) of New World origin, and presumably result from an ancient cross between an Old World A genome and a New World D genome. Euploids (exact multiplication of haploid chromosome number) of these plants have 52 somatic chromosomes, and are frequently designated as AADD. Three additional New World allotetraploids occur in the genus, including *G. tormentosum* from Hawaii, *G. mustelinum* from northeast Brazil and *G. darwinii* from the Galapagos Islands.

Acala Cotton. The Acala type is an upland cotton (*G. hirsutum*) originally selected from germplasm introduced from Mexico to the USA.

Egyptian Cotton. Egyptian cotton belongs to *G. barbadense* and originated from crosses between Peruvian cotton and Sea Island cotton.

Sea Island Cotton. Sea Island cotton also belongs to the *G. barbadense* species and has the best fiber quality in the world. Sea Island cotton originated in the West Indies and Sea Islands of the southeast coast of the USA.

Pima Cotton. Pima was developed from crosses between Sea Island and Egyptian cotton.

Tangüis Cotton. It is a kind of *G. barbadense* cotton grown in Peru that was derived from a cross between sporadic village cotton of Indian tribes and an upland variety.

Nodes

The main stem of the cotton plant comprises many nodes, each capable of producing a branch. Except under very rare conditions, each node above numbers 5-7 produces a fruiting branch. The distance between two nodes is called internodal length. The internodal length is a genotypic character but depends highly on growing conditions. The three main factors responsible for longer internodal length are frequent irrigation/rain, excess supply of nitrogen and inability of the plant to retain fruit (due to heat sterility, insects and other factors).

The internodal length is reduced due to a shortage of water and other stresses. The internodal length starts to increase as the stress subsides. The number of nodes produced decreases with a delay in the sowing season. For physiological assessment and monitoring purposes, nodes are usually numbered from the bottom to the top of the plant

The cotyledonary leaves form the first node on the main stem, which is the only node that never has a branch on it. Generally, under the close spacing conditions of a crop, a number of subsequent nodes (5-6) do not bear branches either, unless the terminal of the plant is damaged.

The first node above the cotyledonary leaves is always longer. The next nodes have almost equal internodal length and then the length starts increasing and ultimately decreasing close to termination of growth.

Branches

Branches on the main stem are located in a spiral order, angled along the main stem. The cotton plant has two types of branches: monopodial and sympodial. Some genotypes may have both monopodial and sympodial branches, while others may have only sympodial branches. Monopodial branches are larger than sympodial branches. The distance between the last monopodial branch and the first sympodial branch is only one internodal length.

Monopodial Branches

The branches that do not bear fruit directly are the monopodial branches. Monopodial branches are also called vegetative branches and are always formed at the base (first few nodes) of the cotton plant.

Genotypes exist in which monopodial branch formation is suppressed. Monopodial branches give the plant a bushy look and usually result in a slow rate of boll formation compared to a sympodial-type plant. In general, a monopodial branch has a higher number of bolls than a sympodial branch.

Plant spacing has a great influence on the number of monopodial branches. Closer spacing reduces (may even eliminate) the appearance of monopodial branches.

Sympodial branches bear fruit directly, so they are called fruiting branches. The secondary branches on monopodial branches are also sympodial and bear fruit directly.

Once a sympodial branch has formed on the main stem, one or two branches are formed on every subsequent node until the plant is physiologically exhausted and growth terminates.

Once a sympodial branch has formed at a main-stem node, the plant is no longer able to produce monopodial branches above that node. The node at which the first sympodial branch will appear is a varietal character, but it is also affected by agronomic practices and treatments.

Most sympodial branches are primary branches, but they may have secondary or tertiary branches. Also, perennial cotton plants may have no primary sympodial branches.

Cultivated varieties have a higher number of sympodial branches than monopodial branches.

Plants with only sympodial branches enter into the fruiting phase of growth earlier than plants/varieties also having monopodial branches.

Flower

The cotton plant has a complete flower, surrounded by bracteoles, with a well developed calyx, corolla, gynoecium (female flower parts) and androecium (male flower parts).

The cotton flower has three bracteoles inserted above the nectaries around the flower base. Bracteole size depends on genotype and ranges from 1-3 inches in length and width among cultivars. Bracteoles have many deep cuts (teeth), are green in color, and do not change their color significantly until the boll is ready to open.

The calyx is cup shaped with five teeth indicating five sepals united into one. The corolla has five large petals tapering toward the bottom.

Petals are showy, white, white-creamy or even rose or mauve in color in *G. hirsutum*. The petals of *G. barbadense* are usually bright yellow. Petals sometimes have a dark rose-colored spot at the base in *G. hirsutum* and almost always in *G. barbadense*. Petals are closed in a whirl until the day of anthesis.

Stigma, style and ovary form the female part of the flower in cotton. The ovary comprises 3-5 carpels, forming the corresponding number of locules in the boll. Each carpel has a number of ovules (depending on genotype) arranged in two vertical rows. The number of fertilized ovules corresponds to the number of seeds in each locule if fertilized ovules were not aborted.

Stamens, which form the male part of the flower, are numerous. The lower parts of the stamen filaments are united in a tube. The upper parts contain anthers bearing pollen grains. Pollen grains are round and have spikes. Their color is creamy or creamy yellowish in *G. hirsutum* and yellow in *G. barbadense*.

The flower bud is the first organ that lays the foundation for lint and seed production on the plant. On average, a flower bud matures into a flower in about 25 days. As the flower readies to open, petals push the bracteoles apart to emerge. Opening or unwhirling of petals is an indication that the flower is ready for dehiscence of anthers and pollination of ovules. Petals continue to grow for 10-25 hours even after the flower has opened but at a much slower rate than before opening. Petals change color to pink on the day of opening. The change in color happens irrespective of the fertilization process. Pollen grains are shed just before or soon after the petals open and expose anthers to direct light. Bracteoles begin to dry after the boll has opened but do not fall off by themselves. Defloration increases the height of the plant.

Cotton pollen grains cannot remain viable for a long time in storage. After twenty-four hours, storage of the open flower in a household refrigerator greatly reduces the germinability of pollen grains. A flower bud collected the day before anthesis retains pollen viability for a longer time (a few days more).

Pollination occurs soon after anthesis, but fertilization occurs 12-20 hours later. Many pollen tubes grow on the stigma, but not all of them reach the ovules. Thus, flower ovules may or may not all become fertilized. At least one ovule must be fertilized to assure retention of a boll. However, sometimes even a minimum number of fertilized ovules per flower may not be enough to hold the boll on the plant. Rainfall in the morning hours at the time of anthesis negatively affects seed setting efficiency because free water ruptures the pollen grains.

Following pollination, the pollen tube growth rate is slow for the first two hours, then increases to a maximum of 3 mm per hour, with pollen tube growth being slowest in the ovary.

Temperature has a critical effect on pollen tube growth compared to light and humidity. The threshold temperature limits for proper pollen tube growth range from 15°C to 50°C.

Root

The cotton plant has a tap root system. The root has the primary function of absorbing and transporting water and nutrients from the soil to the plant parts and anchoring the plant. The radical forms the primary root that grows downward into the soil. The size of the root depends on the physical texture of the soil, soil fertility, soil temperature and the amount of water in the soil. The root can go deeper than three meters. The total root length varies by varieties and species in addition to growing conditions. On average, roots makes up one-fifth of the dry weight of a plant. Small fibrous roots start to die as the plant enters into its heavy boll-load stage and beyond.

Roots grow rapidly during seedling development and nearly attain their maximum length by the time the plant enters the reproductive stage. Root depth is almost six times the height of the shoot at thirty days after planting. At this stage, the lateral growth is only 2.5 to 3 times the height of the plant. In most cultivated varieties, 40% of water needs is absorbed by the top one-quarter of the root system, 30% by the second quarter, 20% by the third quarter and only 10% by the fourth quarter.

Seed

Seed Formation and Germination

Technically, a seed in cotton is a fertilized ovule that may or may not mature into a full grown seed. Seed development can be divided into three stages: enlargement, filling and maturation. Seed development is strongly related to potassium availability. In about three weeks, the fertilized ovule develops the size of a full-grown seed. During the next three weeks, the seed accumulates oil and proteins in the embryo during the filling stage. The maturation stage is characterized by significant physiological processing, including hardening of the seed coat.

Seed size varies greatly among species and varieties. Seed size also varies within a variety. The average length and width of upland cotton seed is about 10 mm x 6 mm. G. *barbadense* seeds are smaller than G. *hirsutum* seeds, and they are mostly naked or have less fuzz. G. *arboreum* and **G. *herbaceum*** seeds are even smaller than G. *barbadense* seeds, but they are fuzzy. Seed index is the weight of 100 fuzzy seeds. G. *hirsutum* varieties have 7,000 to 8,000 seeds per kilogram.

Unfertilized ovules make up the majority of motes, but some fertilized ovules may abort and form larger motes of various sizes.

Seed can be stored for over five years if the moisture level is less than 10% and there is good aeration. The period of safe storage decreases with the increase in seed moisture content. Seed developed during humid conditions is more prone to deterioration.

Seed harvested soon after boll opening has a dormancy period called the "quality conditioning" period. Cottonseed is usually dormant for 3-4 weeks. Abscisic acid inhibits germination and the rapidity of germination. The concentration of abscisic acid, an endogenous (related to the metabolism of nitrogenous compounds) constituent of the developing cottonseed, decreases as seed nears maturity. In addition to abscisic acid, other reasons for seed dormancy are fatty acids, impermeability and hardness of the seed coat, making it unable to absorb water or oxygen.

The appropriate depth for planting cottonseed is 3-4 cm. The germination process does not begin unless moisture is absorbed into the seed. If enough moisture is available, the radical emerges through the micropylar end of the seed in 2-3 days to form a root. Micropyle is a minute hole on the tip of the seed. The hypocotyl is the stem tissue between the radical and cotyledons. As soon as the radical is formed, hypocotyl cells expand and push the cotyledons out of the ground. The germination process is slow under low soil temperatures. For optimum germination under field conditions, soil temperature must be at least 60°F. At least 50-60 heat units are required for a seedling to emerge from the soil. Under optimum conditions in the field, the germination process is completed in about 6-8 days.

Composition

In general, ginned cottonseed is composed of:

Product	Yield
Crude oil	16%
Hull	27%
Meal	46%
Linters	8%
Waste	3%

Seeds of *G. hirsutum* after ginning (fiber removal) are usually covered by short hairs called fuzz. After the lint has been removed from mature seed, manually or mechanically, the remaining cottonseed may have fuzz (short fibers), or it may be without fuzz, usually referred to as "naked." The fuzz removed from the seed is called linters. Linters on the seed can be as high as 10% of the seed weight. Only 1% of the total seed weight of naked-seed varieties is contributed by linters. Linters are categorized as first cut linters, second cut linters, and mill cut or defibrated linters. Linters are about 75-80% cellulose. The degree of fuzziness varies greatly among varieties within the cultivated species.

In most cultivated varieties, the oil content of whole seed ranges from 15-22% of fuzzy seed. The oil content of the kernel ranges from 28-35%. Cultivated tetraploid species have a higher oil content than cultivated diploids. Oil and protein percentages are inherited independently of each other, but there is a negative correlation between oil and protein levels in response to environmental effects. Most oil in the seed accumulates 25-40 days after anthesis.

Breeders seek to eliminate cyclopropenoid fatty acids, but they seek to increase the amount of unsaturated fatty acids (especially oleic and linoleic). The saturated fatty acids should be reduced. In proteins, essential amino acids, especially lysine, methionine and isoleucine, should be increased. Cottonseed oil has a nutritional value of about 10 k cal/g. Average digestibility of cottonseed oil is 97%.

Oil can be extracted either mechanically (screw or hydraulic press) or chemically by adding solvents. Solvent extraction increases the recovery percentage. Cottonseed oil has gossypols that can be neutralized in the extracted oil, but the oil requires bleaching and deodorization.

The shelf life of cottonseed oil is comparable to other oils like groundnut and safflower.

The seed coat (hull) is soft and permeable prior to boll opening. Many layers of seed coat are cemented together immediately prior to and during boll opening to make the seed coat impermeable and hard. The hull is black and has six layers, providing a high level of protection to its embryo compared to most other agricultural crops. It is about 20-50% of the seed weight depending upon the seed size and the thickness of the seed coat.

The seed hull contains about 35-60% alpha cellulose, 19-27% pentosans, 15-20% lignin and 5% ash, proteins, fat, etc.

After the oil has been extracted from the cottonseed, the residue, called cottonseed meal is high in proteins and is usually marketed for animal feed.

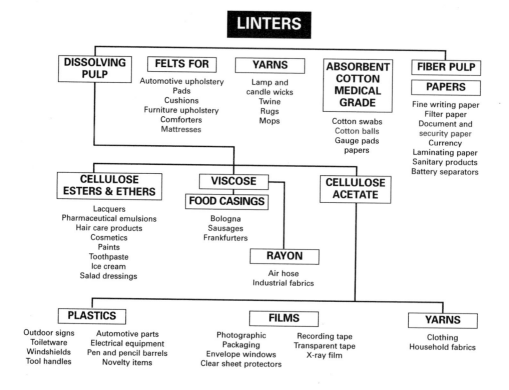

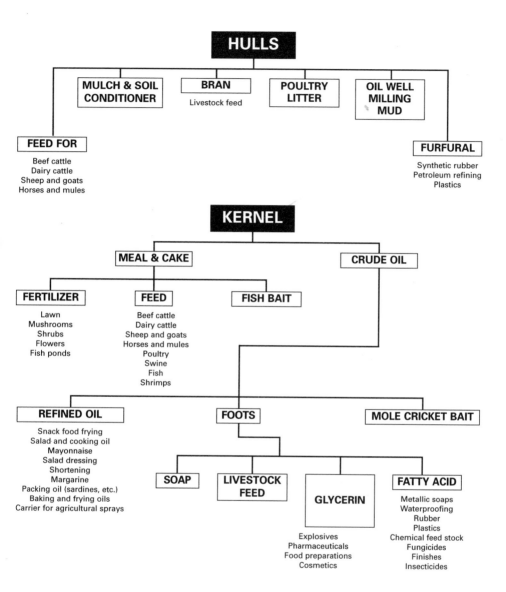

Seed Quality

Cottonseed quality can be affected by pre-harvest and post-harvest operations. The pre-harvest operations include production practices and weather conditions. The effect of meteorological and physiological parameters on seed development is significant and results in different seed quality from different pickings. Seed from the last picking is smaller and less mature than that from earlier pickings.

The post-harvest operations that affect seed quality include harvesting, processing and storage. Mechanical injuries reduce storage life, vigor and emergence of planting seed. Mechanically damaged seeds are liable to infection by microorganisms and may rot in the soil before germinating.

Most of the post-harvest damage to seed is in the form of mechanical injury, though physiological deterioration can also occur due to poor storage conditions. Some varieties with relatively thin seed coats suffer more from mechanical damage than others.

Saw ginning causes more mechanical damage than roller ginning, which causes almost no damage. During acid delinting, acid can be absorbed though injured seed coats, and will burn the embryo, making it unable to germinate, or if it does germinate there will be burned spots on cotyledons.

Seed Delinting

Cottonseed must be able to flow down a chute to have accurate metering during planting operations. The process of removing fuzz from seed is called delinting. Mechanical, flame and chemical operations are used in delinting. Mechanical delinting does not remove all fuzz on the seed and can cause mechanical damage to the seed coat, reducing seed quality. Flame delinting (flame "zipping") removes fuzz better than mechanical delinting but can also reduce seed quality through exposure of the seed to high temperatures.

For chemical delinting, acids have been used in concentrated, diluted and gas forms.

The concentrated acid treatment requires that the seed be washed with water after treatment to remove any remaining sulfuric acid. Fuzz is dissolved in the acid, but disposal of water containing acid is a problem. Seed treated with diluted acid (usually 10% sulfuric acid) does not require washing with water. The acid-treated seed must be centrifuged and neutralized of residual acidity. The removed fuzz can be processed for additional uses. Hydrochloric acid has been used in gas delinting of cottonseed. The treated seed must be scrubbed to remove the fuzz weakened by acidic gas. Acid delinting also kills disease microorganisms on the seed coat, as well as pink bollworm larvae hidden in seeds.

Leaf

The cotton plant has two kinds of leaves: cotyledonary leaves and true leaves. Cotyledonary leaves emerge before the true leaves. The cotton seed has two well-developed cotyledons. The two cotyledons always form the first green leaves (also called cotyledonary leaves or seed leaves). Cotyledonary leaves have a rounded shape and have a short life, shortest among all leaves on the plant.

Cotyledonary leaves are almost always two in number. On very rare occasions an abnormal embryo will have three cotyledons. Cotyledonary leaves appear to be attached to the stem directly opposite each other but in fact one is placed slightly above the other. One cotyledonary leaf falls off before the other with a margin of three to ten days. Cotyledonary leaves have a maximum life of 40 days.

Cotyledonary leaves are thicker than true leaves and do not have pointed edges (lobes) like true leaves. The main stem emerges from the middle of the cotyledonary leaves.

True leaves have pointed edges from the beginning. By the time the first true leaf unfolds, 6-7 other leaves have already been formed. True leaves have 3-5 lobes. The cuts (lobe sinus) may be deep or may only be half cut. Leaves with deep sinuses provide more aeration in the crop canopy and are called "okra type" and "super okra," depending on the size of the sinuses. Okra and super okra leaves, due to deeper cuts, may have a higher perimeter but a lower leaf area index compared to broad leaves.

True leaves reach their functional maturity in about twenty days and then begin to support other growing organs, including bolls. On average, leaves continue this support for the next forty days. Premature yellow leaves are usually half as thick as normal green leaves.

The cotton plant has a spiral phyllotaxy (arrangement of leaves on the main stem). Every leaf is located at a three-eighth turn from the last leaf. The arrangement can be clockwise or counterclockwise in the ultimate ratio of 1:1.

Each leaf axel above node 5-7 has at least two buds, which at different times gives rise to a sympodium. The second bud often produces only a single fruiting point that may or may not abscise, or the bud may fail to break dormancy altogether.

G. barbadense has the biggest leaves among cultivated species. A full-grown leaf is 12-15 cm in length and width. Leaf color in cotton plants may be various shades of green or dark red.

Both sides of the leaf have stomata. The upper surface of the leaf has 2-3 times more stomata than the lower surface, with 100-130 stomata/cm2 on the upper surface and 40-50 stomata/cm2 on the lower surface.

Leaves may have one nectary on the underside of the main rib, or they may be without nectaries. The nectary secretes sugary contents that may contribute to stickiness in cotton and also provides nutrition for multiplication of certain beneficial and harmful insects.

Leaves of most cultivars have hairs. Hair density varies by species and variety; however, 20- to 22-day-old leaves have the highest hair density. Leaf hairs start falling as they age. Higher hair density usually results in higher amounts of trash in harvested lint, particularly in machine-picked cotton. Some bollworms prefer hairy leaves rather than smooth leaf varieties. Hairy varieties are preferred by whitefly. Late sowing reduces leaf hairiness.

Leaves have many functions. Leaves support other organs for various nutritional requirements. The cotton plant absorbs water, nitrogen and other nutrients from the soil. Leaves transform these nutrients from an inorganic form to organic substances through photosynthesis. Leaves store potassium for use during boll maturation when the plant is unable to meet potassium needs from the soil. The plant has the highest need for potassium at the time of boll maturation. Leaves also absorb systemic insecticides and save the plant from pest damage.

The condition of the plant's leaves reveals much about the needs of the plant. Leaves are the strongest indicators of the need for irrigation. They are indicators of pest attack and of the need for insecticide application. Hairy leaves protect against attack by some insects, but are preferred by other insect pests. Leaves also serve as indicators of nutrient deficiency.

Normal leaves have the highest amount of nitrogen among all above-ground parts (excluding seeds) of the plant. With age, the nitrogen level decreases by almost 50%—less than in the seed. The reason for the decrease in nitrogen is its translocation from older to younger tissues.

There may not be bolls on the plant, yet the plant could have normal leaf size and number. Water-stress affects leaf area more than photosynthetic rate. Soil salinity affects photosynthesis and stomatal conductance. Excessive doses of nitrogenous fertilizers increase leaf area but do not add proportionally to the leaf dry weight.

Leaves may exhibit three types of shedding: premature shedding, normal/age shedding and forced shedding. This process is supported by an abscission layer, which is a layer of specialized cells. It is formed between the leaf petiole and the main stem or branch in response to nutrition, water, or some other stress. Abscission occurs when the cells in the layer expand, cell walls dissolve and the mechanical weight of the leaf causes it to fall.

Many factors, including physiological and environmental, could force the leaves to shed before they reach the age of normal shedding. Abscission is the process which causes the plant to shed its leaves. A change in hormone activity is also responsible for leaf shedding. A number of other factors also enhance the formation of the abscission layer. Spraying certain chemicals called defoliants can also enhance formation of the abscission layer. Defoliation is generally a prerequisite for machine harvesting.

Boll

A fertilized flower takes about fifty days to become an open boll. Boll period is the duration from open bloom/flower to open boll. On average, a boll develops to full size 25-30 days after flowering. By then, seeds have also grown to full size. Heat units per day have a significant effect on the time taken to mature. Extremely low temperatures and frost dries green bolls on the plant. The boll period increases with a decrease in temperature. Night temperature is more important than day temperature.

Boll weight is the weight of seedcotton picked from a single, naturally open boll.

A boll may have 3-5 locules. In varieties that have the genetic ability to bear five-locule bolls, the proportion of the bolls increases toward peak flowering and decreases after boll load increases. Five-locule bolls are more liable to be dropped due to stress at an early stage than four-locule bolls. If a plant has five- and four-locule bolls, seeds from the five locules could produce five- and four-locule bolls. Seeds from a four-locule boll in a plant having five- and four-locule bolls may produce four- and five-locule bolls.

The size, shape and smoothness of a boll's surface vary greatly among varieties. *G. hirsutum* has the largest boll among cultivated varieties, followed by *G. barbadense*, *G. arboreum* and *G. herbaceum*. Most cultivated varieties produce the following boll weight:

Species	Range of Boll Weight
G. hirsutum	3.0-10.0 gm
G. barbadense	3.0-5.0 gm
G. arboreum	1.0-4.0 gm
G. herbaceum	1.0-4.0 gm

Boll size is greatly dependent on climatic conditions; mild conditions produce bigger bolls. The shape of a boll ranges from round to oblong and pointed.

"Hard loc" is a condition in which a boll or part of a boll cracks but fails to open fluffy. Fibers from hard locs are weak and immature and seeds do not have a fully developed embryo.

Agricultural crops are grown for a variety of purposes. For example, potatoes are grown for their tubers, sugarcane for the sucrose in its juice, tobacco for its leaves, oranges for their juice, corn for its seed, and cotton only for an outgrowth (trichome) on its seed coat. Cotton is also grown for lint, and thus the quantity of lint produced per unit area is generally referred to as yield. However, for a farmer who sells seed cotton, seed cotton is yield, but for a farmer who sells lint, only lint is yield. Cotton yields are limited by constraints. The nature of constrains may be different in different countries or growing conditions. There are various ways of measuring yield as well, such as biomass, biological and economic yield, genetic potential and recoverable potential.

Biomass is the total dry matter weight produced by the cotton plant. It includes root, stem, leaves, branches, all forms of fruiting parts and seedcotton. Biological yield is the dry matter weight of all above-ground parts of the cotton plant. Biomass less the dry matter weight of root is equal to biological yield.

Economic yield in any crop/plant is the ultimate product for which a crop is grown. In the case of cotton, it is primarily lint.

Genetic potential is an arithmetic calculation of yield based on the genes present in the genotype. The genetic potential of cotton varieties is not known.

The recoverable potential is an upper limit yield that can be realized under a particular set of growing conditions. A genotype can exhibit a higher recoverable potential under favorable weather conditions, while adverse conditions can inhibit the plant from expressing its genetic potential for economic yield. Thus, genotypic interaction with growing conditions determines the recoverable potential. The recoverable potential is always less than the genetic potential in cotton. The difference will be higher under unfavorable conditions and lower under conditions that are favorable for a genotype.

Earliness in cotton is defined in many ways. It is defined as appearance of the first flower, early fruit formation/retention, early boll opening, termination of crop, seedcotton picked in the first pick compared to the total harvest, and completion date of final picking.

The cotton plant, being indeterminate in nature, supports vegetative and reproductive growth at the same time during most of its life cycle.

Production Technology

Production technology consists of a set of recommended practices for given growing conditions. Recommended practices could have short-term and/or long-term effects. The three fundamentals of a successful production technology are maximum yield, best fiber quality and low cost of production. Cotton responds quickly to growing conditions and input applications. The response depends on the intensity of changes in growing conditions or input applications.

Growing conditions differ significantly among countries, among regions within a country, among farmers within a region and among fields at the same farm. Consequently, no single set of recommendations can be applied universally. Further, recommendations change with changes in crop and weather conditions.

Production technology can be low input technology, usually resulting in lower yields, or high input technology utilizing intensive input applications and usually resulting in high yields. Production technology in cotton can be tailored to achieve a specific yield level. So the technology could be a low yield technology or a high yield technology.

A given set of production practices has many components. Technology includes fertilizer, water, insecticides and other chemicals/inputs, as well as the selection of the variety to be grown. Thus, technology is the sum of interactions among all inputs, and a decision as to when and how to apply each input determines whether a particular approach is successful. Over- and under-use of one input can significantly change the quantity and timing of other inputs.

Current production technology in cotton relies heavily on pesticide use, which affects the sustainability of cotton production worldwide. The success of cotton production lies in the least use of chemicals, particularly those that are harmful to the environment. Best use of available resources, like water and soil, based on their long-term use, is an important determinant of the sustainability of cotton production.

Organic Cotton

Growing awareness of the environmental impacts of pesticide use has led to a growing demand for organic cotton. Organic cotton is produced without the use of synthetic chemical pesticides, fertilizers, growth regulators, hormones, defoliants or any other chemicals that may add to the environmental load. It is not just the elimination of chemicals and fertilizers, but a system of production and processing that seeks to maintain and replenish soil fertility and the ecological environment of the crop.

Organic cotton may or may not be certified. In order for the marketer to claim the cotton to be organic, it must be certified by a designated certifying organization. The certifying organizations/companies have established various standards that must be met to have produce certified as organic. In order for cotton to be certified as organic, the conditions of a transitional period varying from 2-3 years must be met. The cotton produced during the period when chemicals are not used, but before certification as organic cotton, is called transitional cotton.

Precision Agriculture

Another response to the problem of intensive pesticide and fertilizer use is precision agriculture, which is the management of variable rates of application of inputs. The idea of precision agriculture is implemented through site-specific crop management that matches resource application and agronomic practices with soil attributes and crop requirements as they vary across fields and farms. Precision agriculture can also be referred to as differential treatment of field variation, in contrast to the traditional uniform treatment of all fields on a farm or all farms in an area.

2 *Breeding*

Plant breeding is the art and science of changing and improving the heredity (genetic abilities) and performance of plants. Breeding can also be defined as the use of techniques involving crossing plants to produce varieties with particular characteristics (traits), which are carried in the genes of the plants and passed on to future generations. Breeding can also be defined in many other ways. Breeding is an application of genetic principles for the improvement of plants and other organisms.

Conventional/traditional plant breeding refers to techniques other than modern biotechnology, in particular cross-breeding, back-crossing, etc. In practice, breeding in cotton and other crops generally refers to development of new, superior varieties. Other and more recent techniques used in breeding include state-of-the-art breeding methods such as genomics, marker assisted breeding (MAB), biochemistry and cell biology.

Cotton is generally self-pollinating, but in the presence of suitable insect pollinators can exhibit some cross-pollination. Cotton is classified as an often cross-pollinated crop but for breeding purposes, it is treated as a self-pollinated crop, which is true for all cultivated species. Cotton, in spite of being an often cross-pollinated crop, does not suffer from in-breeding depression (loss due to self-pollination).

The extent of natural out-crossing in cotton depends on the climatic conditions where cotton is grown. The extent of natural cross-pollination varies even within a country. The cotton pollen grains cannot be carried by wind, and only insects carry pollen from one flower to another.

Genetics is a science of heredity. It is also a science of similarities and differences. This is a science that tells how traits are inherited and why an offspring is similar or different from the parents. Gregor Mendel published his work, *Experiments with Plant Hybrids*, in 1856. His work was so brilliant and unprecedented at the time it appeared that it took 34 years for the rest of the scientific community to catch up to it. Mendel's work was rediscovered in 1900 and the science of genetics was born.

Genetic mutations are the changes between or within chromosomes that may alter a phenotype, giving it greater or less advantage in the process of natural selection. Mutation brings a permanent change in the DNA sequence, which could have a minor or major phenotypic effect. Mutations in the reproductive cells are heritable and transmitted to the next generation. Mutations in the somatic (vegetative) cells are not heritable but may be transmitted once to daughter cells.

Mutagens are agents, including chemicals and radiations, used to create artificial mutations. Mutation was introduced by X-irradiation in 1927.

Mutation breeding is the use of mutagens, both physical and chemical or in combination, for realizing new variability. Ionizing radiations (radiation capable of creating ions) such as gamma rays have been used in cotton for inducing sudden and often drastic changes in many instances. Most mutations are recessive and lethal in nature. Cotton being an allotetraploid has less chance of mutation expression. Mutation breeding is not commonly used in cotton now, but varieties have been developed using mutation and adopted on a commercial scale in some countries. Mutation breeding was mostly tried in cotton during the 1960s and 1970s with the objective of creating non-existing characters. Chemicals and ionizing radiations were used to create permanent changes in the existing genomes in many countries. The most significant challenge in mutation breeding lies in the detection of a desirable mutation that is not linked to any negative effect.

Epistasis is the interaction between non-allelic genes where presence of one allele (epistatic) at one locations prevent the expression of another allele at another location.

Pleiotropic effect is the ability of a gene to affect more than one phenotypic character.

Breeding Methods

The most commonly used breeding methods other than mutation breeding that are applied to cotton are: introduction, selection, hybridization, and conventional and traditional breeding. Introduction is the direct adoption of native/developed germplasm from elsewhere. Selection is the process of planned improvement in the performance of specific cultivars for certain traits through conscious choice. The sources of variation may be natural mutation, segregation within a population and natural out-crossing. Commonly used selection methods in handling the segregating population developed through hybridization are pedigree, bulk and mass selections.

Hybridization is the crossing of genetically different parents for the sake of creating variability, often with the purpose of obtaining genotypes with transgressive performance. Hybridization (crossing between two parents) results in new combinations, but drastic changes should not be expected. This is the most widely used method of developing new cotton varieties. Conventional Breeding/Traditional Breeding is the application of introduction, selection and hybridization methods for developing or improving genotypes/varieties.

F1. The first filial generation, produced by crossing two genotypes, is called the F1 generation, usually created by a single cross involving two pure/homozygous genotypes. If the parents are homozygous and pure, the F1 generation plants should be all heterozygous but, phenotypically, all plants within a cross should be similar. Variation within the F1 population is proof that the parents were not pure and homozygous. The F1 generation exhibits the maximum hybrid vigor.

F2. The F2 generation is the second filial generation produced by planting F1 generation seed. Segregation of traits begins in the F2 generation. The F2 generation expresses maximum variability within a cross.

The number of generations required to achieve homozygosity in cotton depends on the diversity of the parents involved. In general, by F6 most characters are believed to be fixed and homozygous (true breeding) in cotton, provided there was no out-crossing in the segregating populations. Selections made from the F6 generation should breed true.

Below are a number of commonly used terms in plant breeding.

Monogenic. A character that is controlled by a single gene is called a monogenic character.

Double Dominant or Double Recessive. A character that is controlled by only two pairs of dominant or two pairs of recessive genes is called double dominant or double recessive, respectively.

Single Cross. A cross between two true breeding parental lines or inbreds.

Double Cross. A cross between two F1 hybrids.

Three-way Cross. A cross involving one F1 hybrid as a parent and one new parent not used in the first combination.

Wide Cross. Refers to the plant breeding technology/technique utilized to cross two plant species that would not normally cross in nature.

Back Cross. Crossing of the F1 generation with one of the parents used in the original cross. The back-cross technique is mainly used to improve a specific trait or a limited number of traits in a desirable variety. Usually 4-6 back-crosses need to be performed for retaining or obtaining the maximum improvement from a desirable variety.

Recurrent Parent. A desirable variety that is used again and again in back-crossing for retaining the best characteristics out of that variety is called a recurrent parent.

Pedigree Selection/Progeny Row Selection. The hybrid population in the F2 generation is subjected to the selection of single plants, which are grown in individual rows in the F3 generation. The best performing lines in the F3 generation are selected within selected lines. Such a procedure is repeated for 3-4 generations until the material is believed to be homozygous based on morphological performance. Such a process, where historical performance of selected plants and progenies can be traced, is called pedigree selection or progeny row selection. The process is labor intensive but more reliable.

Bulk Selection. In the bulk selection method, the F2 population is grown without selection and bulk seedcotton is harvested. A sample from the bulk seed is planted in the F3 generation and repeated until (F5 or F6) plants within a bulk are believed to be homozygous. Single plant selections are made in the homozygous but heterogeneous population (all plants similar but will segregate in the next generation). Bulk selection is a simple process but carries the risk of losing good plants in the process as only a portion of the seed is planted in the next generation.

Mass Selection. The mass selection method is selecting many plants at the same time. Mass selection is not a procedure to deal with the segregating material; rather it is usually applied to seed production. Many true-to-variety plants are selected and bulked for maintaining the varietal purity.

Selection Coefficient. Selection coefficient is a measure of the rate of transmission through successive generations of a given allele compared to the rate of transmission of another allele.

Natural Selection. Natural selection is the process by which inheritance needs and abilities of a genotype are more or less closely matched to resources available in their environment, giving those with greater fitness a better chance of survival and reproduction.

Selfing. Selfing is the process of covering the flower bud to ensure self-pollination. Paper clips can be used to seal the cotton flower petals before they open and expose stigma for cross pollination. The cotton flower can also be covered with a paper bag to ensure self-pollination, or even tied with a thread.

Heritability. Heritability is the ability of a character to be transmitted from a parent to the hybrid. A character may have a high or a low heritability. High heritability means that there are more chances of a character to be transmitted to the next generation.

Cultivar. An agriculturally derived variety of a plant that can be distinguished from a natural variety. A cultivar can also be defined as a variety of plants produced through selective breeding by humans and maintained by cultivation. Cultivar is also a group of plants that have no known wild ancestor.

Genotype. The set of alleles that an individual plant possesses. The set of alleles could be homozygous or heterozygous. All homozygous alleles will breed true, while the heterozygous alleles will produce a segregating population.

Phenotype. A seeable physical and biochemical characteristic of an organism. A phenotype can also be the outcome of a genotype interaction with the environment. The same cotton genotype may give a different phenotypic expression under different growing conditions.

Strain. A genotype that has been genetically purified and tested for years for its yield performance and fiber quality and is considered to be a candidate for release as a commercial variety. Although a number of genotypes reach the strain stage, only a few, or sometimes none of them, are approved as a variety.

Variety. A different kind, sort or form of the same general population. In agriculture, a group of homozygous similar-behaving plants is called a variety. In practice, variety usually means a strain that has been officially approved for commercial production.

The variety approval process varies among countries. The process ranges from strict government evaluation and approval through different layers of consideration, to a no-approval process. No-approval refers to private companies who have established their own standards and are not required under local laws to test their varieties against other varieties and follow an approval process. The formal variety approval process is difficult in countries where more than one institution/agency is involved in developing cotton varieties and they compete against each other.

Cultivar vs. Variety. Cultivar and variety are often inadvertently used as synonyms. "Cultivar" is the accepted name used by the scientific community for a cultivated variety of cotton. "Variety" is typically used in common language for the same identity. The scientific community discourages the use of "variety" for a cultivated plant because the term has a different meaning with respect to plant taxonomy.

Germplasm. Living tissue from which new plants can be grown. It could be a seed or a plant part such as a leaf, a piece of stem, pollen or just a few cells that can be cultured into a whole plant. Germplasm contains the genetic information for the plant's heredity makeup. In cotton, germplasm refers to the stock of genotypes preserved for study and utilization in breeding. Cotton germplasm is stored as seed.

HYBRID COTTON

Heterosis refers to the hybrid vigor or better performance of the hybrid than the mean of parents, usually expressed in terms of percentage. The maximum hybrid vigor in cotton is realized in the F1 generation. Shull coined the term heterosis in 1914.

Heterobeltiosis is the increase in vigor for any character over the best parent. Heterosis and heterobeltiosis may not be of any use unless the F1 also surpasses the commercial variety. Superiority/hybrid vigor over the commercial variety (or varieties) in the area is called *economic* or *useful heterosis*. In cotton, lint is the yield and vegetative superiority (pseudoheterosis) is not desirable. Euheterosis or true heterosis is the real objective in cotton.

Transgressive inheritance of a particular character is the performance of a hybrid better than either of the parents for that particular character.

In cotton, those hybrids that are better (in yield and/or quality) than the varieties used in commercial production in the area are commonly called hybrid cotton. India was the first country in the world to start commercial production of hybrid cotton. H-4, the first intrahirsutum hybrid, was released in 1970.

Two important aspects of hybrid cotton are 1) identification of good combiners which, when crossed, produce a hybrid that gives acceptable higher yield over commercially grown varieties, and 2) economical production of hybrid seed for commercial planting.

Production of hybrid seed can be accomplished in several ways. Hand emasculation and pollination are expensive ways to produce hybrid seed. Average seed setting efficiency in hand emasculated and pollinated flowers is always low, usually around 50%. Genetic Male Sterility (GMS), Cytoplasmic Male Sterility (CMS) systems, and gametocides can be employed in large scale production of commercial cotton hybrid seed.

GMS is a genetic inability of the cotton plant to produce fertile pollen grains. The first genes known to condition such sterility were the double recessive ms5 and ms6. GMS lines having ms_5 and ms_6 genes normally segregate into 1:1 ratio of a male sterile–male fertile plant when the $ms_5ms_5\ ms_6ms_6$ is pollinated from the sister $Ms_5ms_5\ ms_6ms_6$ fertile plant. Many other genes responsible for genetic male sterility have been identified.

Converting a Genotype into a GMS Line

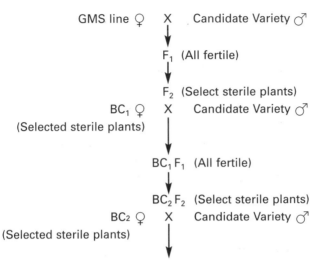

GMS line ♀   X   Candidate Variety ♂

F₁ (All fertile)

F₂ (Select sterile plants)

BC₁ ♀   X   Candidate Variety ♂
(Selected sterile plants)

BC₁F₁ (All fertile)

BC₂F₂ (Select sterile plants)

BC₂ ♀   X   Candidate Variety ♂
(Selected sterile plants)

Most varieties will produce ¹/₁₆ sterile plants in F₂ but the proportion of sterile plants will continue to increase in the back-cross F₂s.

Continue back crossing to BC₄ or BC₅.

Select sterile plants in F₂ of BC₄ or BC₅ and pollinate them with sister fertile plants resembling the candidate variety. Keep record of the male plants. The F₁ segregation into 1:1 ratio is an indication that the conversion is complete.

Maintain the new GMS line by sib pollination with fertile plants.

Conversion of a genotype into a GMS line is complicated and lengthy compared to conversion into a CMS line. A shortcut method is available where the selected sterile F₂ plants have to be crossed with F₁ plants, but it is even more complicated and requires a great deal of care. The main advantage of the GMS system is that a normal male parent is able to restore fertility in F₁ crosses.

Cytoplasmic Male Sterility (CMS) was developed by Vesta Meyer in the USA by transferring the *G. hirsutum* (AD) genome into the cytoplasm of *G. harknessii* (D). Any variety can be converted into a CMS line by following a simple back-cross technique. A converted CMS line is used as a recurrent male parent. A CMS line is also called an "A line." The fertile isogenic line of an A line is called a "B line." A CMS line is maintained by pollinating an A line with the respective B line. The progeny will be sterile and cross-pollination is required for every generation.

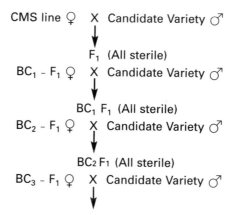

Converting a Genotype into a CMS Line

CMS line ♀ X Candidate Variety ♂
↓
F_1 (All sterile)

BC_1 - F_1 ♀ X Candidate Variety ♂
↓
BC_1 F_1 (All sterile)

BC_2 - F_1 ♀ X Candidate Variety ♂
↓
BC_2 F_1 (All sterile)

BC_3 - F_1 ♀ X Candidate Variety ♂
↓

All F_1 plants including the back crosses will be sterile but plants having more similarity to the candidate variety be selected for back crossing.

Repeat back crosses to BC_4 or BC_5 until the sterile plants look similar to the candidate variety.

Maintain the new CMS line (A line) by pollinating with its respective male parent (B line).

CMS lines without cross pollination are always taller and more vegetative compared to their respective B lines. CMS lines exhaust physiologically much later compared to the respective B lines. Without boll set, CMS lines continue to bear flowers even after the respective B lines have terminated.

A genotype that is able to restore fertility in a CMS line is called a restorer or "R line." A normal variety has to have a restorer gene (genes) to induce fertility in the F1 that is produced by crossing on to a CMS line. Restorer fertilized CMS lines produce fertile progenies. Any genotype can be converted into a restorer line by either keeping the restorer line as a female parent or as a male parent for using *G. harknessii* cytoplasm or *G. hirsutum* cytoplasm, respectively.

It is safer to use *G. harknessi* cytoplasm for the restorer strain. In each case 4-5 back-crosses will be required until a new genotype is converted into a restorer line. A test-cross should be made after each back-cross to determine which individual plants carry better restorer genes. In order to confirm the fertility restoration ability of the new R line, it should be crossed with a CMS line and fertility should be checked in the F_1 generation. If all plants are fertile, the line was successfully converted into an R line.

Spraying certain chemicals called gametocides on the plant can also induce male sterility. Such sterility is not transmitted to the next generation. Sterile plants can be pollinated with pollen from regular untreated plants by hand or by insects.

BIOTECHNOLOGY

Biotechnology is a broad term encompassing utilization of living organisms for the improvement of living organisms. Biotechnology is also defined as procedures involving the application of in-vitro nucleic acid techniques and cell and tissue culture techniques to reach an objective, including genetic engineering. Biopesticides are biotech products, but they may not be, and mostly are not, genetically engineered.

Genetic Engineering

The technology for dealing with DNA has become so powerful that it is now routine to construct novel DNA molecules by joining sequences from quite different sources. The product is often described as recombinant DNA, and the techniques (more colloquially) as genetic engineering. The technology is applicable equally to prokaryotes (cells with no definitive nucleus wall) and eukaryotes (cells with a definitive nucleus wall), although the power of this approach is especially evident with eukaryotic genomes. Genetic engineering is one process used in biotechnology. Using the technique of "gene splicing" or "recombinant DNA technology" (rDNA), scientists can add new genetic material to an organism to have it produce a new protein. The protein may create new traits, such as immunity against insects or herbicide chemicals, or may strengthen/improve existing traits. Below are a number of terms frequently used in the field of genetic engineering.

Agrobacterium. The process by which some bacteria can transfer DNA from an endogenous plasmid into cotton and other plants/organisms. Some plant genotypes are more receptive to agrobacterium-mediated transformation than others.

Allele. An allele is one of two or more alternate forms of a gene that may occupy a particular location on a chromosome.

Amplified Fragment Length Polymorphism (AFLP). Amplified Fragment Length Polymorphism is a type of "DNA marker" employed in "genetic mapping" techniques. The technique is specific and repeatable in that it uses restriction enzymes to cleave specific base (nucleotide) sequences in the genomic DNA (from an organism), then uses primers to amplify the resulting DNA fragments by PCR. Since the specific sequences of DNA often differ between species, strain, variety, or individuals (due to DNA polymorphism), AFLP can be utilized to "map" those DNA molecules (e.g., to assist and speed up plant breeding programs).

Bacillus thuringiensis – Bt. Bacillus thuringiensis is a common rod-shaped bacterium occurring in the soil. It has wide distribution in the world and is capable of producing "cry" proteins. These proteins are cleaved in the mid-gut of many lepidopteran insects to produce protein subunits that "bind" to receptors in the insects' gut (i.e., stomach) lining (epithelium cell) and cause membrane disruption. The "cry" proteins are toxic to certain types of insects (e.g., moths such as bollworms) that attack cotton, and the action is specific to those insects. Over 20,000 strains of Bacillus thuringiensis are known.

The target insect must ingest the *Bacillus thuringiensis* "cry" protein for the protein to be effective, otherwise the protein will not work and the target insects will not be affected.

The Bt protein is biodegradable and effective against only those lepidopteron insects that have receptors in the mid-gut that are specific for the particular Bt protein. Any material that can be broken down by biological action (e.g., dissimilation, digestion, denitrification, etc.) is described as biodegradable.

Baculovirus. A class of virus that infects lepidoptera insects. Baculoviruses can be modified via genetic engineering to insert new genes into the larvae, causing those larvae to produce lethal proteins. Many of the harmful insects that damage cotton are lepidopterans.

Bases. Bases are chemical units (adenine A, thymine T, guanine G, and cytosine C) in a DNA molecule that constitutes the genetic code.

Bioassay. The determination of relative strength or bioactivity of a substance based on response of a living organism or tissue. A biological system (such as living cells, organs, tissues, or whole animals) is exposed to the substance in question and the effect on the living test system is measured.

Biolistic® Gene Gun. The word "biolistic" comes from the words "biological" and "ballistic" (pertaining to a projectile fired from a gun) used to shoot pellets that are coated with genes into living tissues (e.g., egg cells or plant tissues) in order to get the recovered organism to express the new genes.

Chromosome. Chromosomes are the location of most hereditary (genetic) material within a cell. Genes are arranged in a linear sequence along chromosomes.

Cloning vector. Cloning vector is a genetic element, usually a bacteriophage or plasmid, that is used to carry a fragment of DNA into a recipient bacterial cell for the purpose of gene cloning.

DNA. DNA (deoxyribonucleic acid) is the substance cotton (or any other organism) uses to store genetic information needed to conduct the process of life. It is a material that encodes the inheritance of characters from parents to offspring. DNA consists of two long chains of nucleotides (base + deoxyribose + phosphate). The two chains are held together in a double helix by hydrogen bonds between base pairs of A:T and G:C along the two chains. The sequence order of the nucleotides in the DNA determines the genetic code for the sequence of amino acids that makes up the proteins of an organism and that controls the expression of the coded genes.

Watson and Crick modeled the structure of DNA in 1953.

DNA Fingerprinting. The process of developing a pattern of DNA fragments that is unique to an individual organism, variety, population, etc. is called DNA fingerprinting. The pattern can be generated by a number of techniques including, but not limited to, digestion of the DNA with restriction enzymes or Polymerase Chain Reaction (PCR) using a series of specific or random primers (very short sequences of DNA that bind to complementary sequences on the target DNA).

DNA library. A collection of cloned DNA molecules (in bacteria as plasmids or in bacteriophage) representing either an entire genome (genomic library) or DNA copies of the mRNA produced by a tissue (cDNA library). A genomic library will contain most or all the DNA of an organism, whereas a cDNA library will contain only the gene transcripts that were being expressed in the tissue at the time the mRNA was isolated. Thus, leaf and root cDNA libraries would differ, as would leaf cDNA libraries from water-stressed and non-stressed plants.

DNA Sequence. The order of nucleotide bases in a DNA molecule is called the DNA sequence.

DNA Sequencing. DNA sequencing is the process of determining the sequence of bases in a molecule of DNA.

Enzyme-linked Immunosorbent Assay (ELISA). Enzyme-linked Immunosorbent Assay is a highly sensitive technique for detecting and measuring specific proteins.

Gene. The portion of a DNA molecule that contains the basic functional unit of heredity is called a gene.

Gene Cloning. A clone is defined as an exact copy of a molecule, cell or organism. Cloning of a fragment of DNA (gene) implies production of large numbers of exact copies of the target gene. The process usually involves splicing the fragment into a plasmid (non-chromosomal, circular DNA genetic units in many species of bacteria that replicate when the bacteria divide), transforming bacteria with the plasmid, and allowing the bacteria to divide many times. In this way millions of organisms are obtained, each with one or more copies of the plasmid containing the DNA fragment.

Genetic Code. The genetic code is the information contained in the DNA sequence that determines the amino acid sequence in protein synthesis. The genetic code is read in sets of three nucleotides or triplets called codons.

Genetic Marker. Any character that acts as a signpost or signals the presence or location of a gene or heredity characteristic in a genotype is called a genetic marker. If the character is a DNA sequence or location, it is called a molecular marker.

Molecular Marker Assisted Selection is the use of molecular markers to follow the inheritance of specific genes and/or traits. Given that selection in breeding is based on the phenotypic expression of the plant, molecular marker assisted selection, regardless of the environmental effect, provides information on which genes are present or inherited from the parents.

Genetically Modified Organism (GMO). Genetically engineered organisms commonly, albeit incorrectly, are called genetically modified organisms.

Genome. The sum of genetic material in the chromosomes of a particular organism. The chloroplasts and mitochondria of a plant each have their own genome of DNA.

Genomics. The scientific study of genes and their role in an organism's structure, growth, health, etc.

Introgression. Transfer of segments of a genome and genes from a donor organism into a different species, subspecies or population.

Locus. The position on a chromosome at which the gene for a particular trait resides is called a locus. Only one gene can occupy a locus, and in any given organism only one allele of the gene can occupy the site on one chromosome. The same or a different allele may occupy the locus on the homologous chromosome (a true copy of the pairing chromosome) of the organism.

Nucleotide. A nucleotide is a sub-unit of nucleic acid (DNA or RNA). Each nucleotide consists of a nitrogenous base (adenine, thymine, cytocine, or guanine), a sugar molecule (deoxyribose in DNA and ribose in RNA) and a phosphate group.

Polymerase Chain Reaction (PCR). Polymerase Chain Reaction is an in vitro method of nucleic acid synthesis by which a particular segment of DNA can be specifically replicated. The method uses thermostable DNA polymerase and amplifies the DNA segment flanked by two primed sequences by repetitive cycles of template denaturation, primer annealing, and DNA replication. Newly synthesized strands can act as additional templates for subsequent cycles, resulting in exponential amplification. It is a rapid, inexpensive and simple means of producing relatively large numbers of copies of DNA molecules from minute quantities of source DNA material—even when the source DNA is of relatively poor quality. PCR can be used for detection of pathogens (including viruses) in cotton, detection of plant viruses in their insect vectors, detection of genetic variability of plant viruses, identification of disease resistance genes in cotton, and for various other objectives involving DNA identification or manipulation.

Polymorphism. Polymorphism is a naturally occurring or induced variation in the homologous or homoeologous DNA sequences of two organisms.

Primers. A primer is a short segment of nucleotides that is complementary to a corresponding short sequence of DNA. Before new DNA can be synthesized the DNA-polymerase must be "primed," that is, provided with a 3'phosphate nucleotide upon which to add new nucleotides. Thus, in PCR, primers are annealed to the denatured DNA template to provide an initiation site for the elongation of the new DNA molecule by the polymerase. Primers can either be specific to a particular DNA nucleotide sequence or they can be "universal." However, the term "universal primer" has various connotations related to procedural objectives. Generally, primers are complementary to nucleotide sequences that 1) may be very common in a particular set of DNA molecules, 2) are highly conserved across different life forms, or 3) have been introduced as a convenience in a molecular cloning technique, e.g., the M13 sequence.

Promoter. A promoter is a region of DNA that lies "upstream" of the transcriptional initiation site of a gene. The promoter controls where (e.g., which portion of a plant), when (e.g., which stage in the lifetime of an organism) and under what conditions (environment) the gene is expressed.

Protein. A protein is a polymeric molecule consisting of amino acids linked by peptide bonds. Proteins are the products of gene expression and are the functional and structural components of cells.

Quantitative Trait Loci (QTL). Specific areas on genomic DNA or a chromosome, usually identified by molecular markers, that influence the level of expression of a multigenic (i.e., quantitative) trait.

Random Amplified Polymorphic DNA (RAPD). Random Amplified Polymorphic DNA is the term used to describe differences in DNA fragments between two or more organisms that result when the genomic DNA is subjected to PCR primed by short, random-sequence primers usually 10 nucleotides in length. The limited number of fragments resulting from one 10-mer primer allows direct observation after electrophoretic separation on agarose gels. Most RAPDs are dominant so heterozygous F2 plants cannot be distinguished from homozygous plants. Detection and reproducibility of RAPDs is strongly influenced by the experimental conditions of the PCR reaction and the template DNA being used.

RAPD can be used in cotton for numerous objectives among which are gene tagging, gene isolation, varietal identification, cultivar purity testing, hybridity determination, population genetic studies, monitoring of gene introgression, measurement of genetic diversity, and recombinant DNA technology.

Restriction Fragment Length Polymorphism (RFLP). Restriction Fragment Length Polymorphism is the term used to describe differences in DNA fragment lengths (sizes) between two or more organisms resulting from digestion of genomic DNA with a restriction enzyme. Fragments are usually identified after electrophoretic separation by "Southern Hybridization" with specifically labeled probes, such as cDNA clones. "Genetic mapping" techniques can use RFLPs because the different sized fragments segregate normally, are inherited as alleles of a locus, and are co-dominant. Since different species, strains, varieties, or individuals may have DNA polymorphisms, RFLP can be utilized to "map" their genome for plant breeding and various other purposes.

RFLP can be used in cotton for cultivar identification, testing and assurance of cultivar purity, validation of pedigree, measurement of genetic diversity and/or similarities, identification of QTLs, marker-assisted introgression and protection of proprietary rights, among other uses.

Ribonucleic Acid (RNA). RNA is a nucleic acid, usually single-stranded, consisting of repeating nucleotide units containing four kinds of heterocyclic organic bases (adenine, cytosine, guanine, and uracil), and the sugar ribose linked into a chain by phosphate ester bonds. RNA has a major function in protein synthesis within a cell. In addition, it is involved in various ways in the processes of statement and repression of hereditary information. The three main functionally distinct varieties of RNA molecules are: 1) messenger RNA (mRNA) which is involved in the transmission of genetic information from DNA to the cell synthetic machinery, 2) ribosomal RNA (rRNA) which makes up a part of the physical machinery of the protein synthetic process, and 3) transfer RNA (tRNA) which is involved in the translation of the genetic code into an amino acid sequence during protein synthesis. A variety of less known riboprotein complexes involved in processing of genetic information are also present in cells.

Technology Protection System. A method of modifying plants, also called "terminator technology," so that the seed produced does not germinate, consequently it is useless to save as planting seed for the next crop. Nutritional and other qualities of the seed are not affected.

Transgene. A "package" of genetic material (i.e., DNA) that is inserted into the genome of another genotype via gene splicing techniques.

Transgenic Cotton. A cotton genotype having genetic material from a non-cotton species is called transgenic cotton.

Transformation. Stable introduction of foreign DNA into plant cells is called transformation. Several methods have been used to introduce foreign DNA into cotton for improvement including:

Agrobacterium-mediated transformation

Direct delivery of DNA via Electroporation

Microinjection

Gene gun

Virus mediated transformation

Vector. In genetic engineering, a vector is a self-replicating agent used to transfer foreign DNA into a host cell.

Milestones in Genetic Engineering of Cotton

1970 Arber and Smith discover restriction enzymes (or endonucleases) that could cut DNA only at specific sequences.

1973 Researchers develop the ability to isolate genes. Specific gene codes are identified for specific proteins.

1980 Scientists discover how to transfer pieces of genetic information from one organism to another.

1982 The first commercial application of this technology is used to develop human insulin for the treatment of diabetes.

1983 The first transgenic plant, a tobacco plant resistant to an antibiotic, is developed.

1985 Genetically engineered plants resistant to insects, viruses, and bacteria are field tested for the first time.

1987 A gene conferring glyphosate resistance is transformed into cotton for the first time.

1988 The first glyphosate-resistant transgenic cotton plant is created.

1989 The first transgenic cotton with a Bt gene expressing the protein at economically viable levels is developed.

1990 After approval from the U.S. Department of Agriculture, multi-location field testing of Bt cotton is started in the USA.

1995 Seed multiplication aimed at commercial planting is started.

1996 Bt (Bacillus thuringiensis) cotton with the Bollgard gene (called Ingard in Australia) is planted on a commercial scale in Australia and the USA.

1997	Stacked gene varieties having herbicide resistance and the Bt gene are introduced.
2000	The 2nd generation of a Bt gene called Bollgard II becomes available.
2003	The Bollgard II gene is grown on a commercial scale for the first time.

NATURAL OUTCROSSING

Cultivated species of cotton are bred as self-pollinated crops, but at least some natural out-crossing occurs under most conditions. Out-crossing is limited in cotton due to two main reasons. In most cultivated varieties, before flowers open and expose stigma for cross pollination, self pollination has already occurred. Cotton flowers open in the morning usually for a few hours after sunrise and by then anther dehiscence has already taken place. As temperatures descend at the end of summer and early fall, anther dehiscence time is delayed to around noon and even later, depending upon the temperature. Extremely low temperatures may prevent anthers from dehiscing. Also, cotton pollen grains cannot be blown by the wind. Pollen grains have to be carried by insects, which is also restricted due to insecticide use in cotton.

Propagation by vegetative material is not a common method of cotton reproduction. Physical safeguards that inhibit the movement of vegetative material from the area should be adequate to prevent gene movement by this means.

Gossypium tomentosum is pollinated by insects. The stigma in *G. tomentosum* is elongated, and the plant seems incapable of self-pollination until acted upon by an insect pollinator. The flowers are also unusual because they stay open at night; most Gossypium flowers are ephemeral: they open in the morning and wither at the end of the same day. The genetic material of *G. hirsutum* may escape from an area or country by vegetative material, by seed, or by pollen grains.

Cotton pollen grains are not translocated by the wind. The only sources of natural cross pollination are insects that carry pollen grains from one flower to the other. Out-crossing in cotton depends on insect activity, climatic conditions and the distance between the two parents.

Movement of the genetic material by pollen is possible only to those plants with the proper chromosomal type, tetraploid AADD genome to tetraploid AADD genome. Among cultivated autotetraploids, this will include *G. hirsutum* to *G. barbadense* or *G. barbadense* to *G. hirsutum*. Among cultivated diploid, it will be to and from *G. arboreum* and *G. herbaceum*.

Natural out-crossing can be avoided by separating two varieties at a certain distance from each other. The distance would depend on the percentage of natural out-crossing occurring under normal conditions. Diploid species like *G. arboreum* and *G. herbaceum* can be used as a barrier between two upland varieties to avoid natural out-crossing.

GOSSYPOL

Gossypol is a pigment produced in glands, seeds, and other parts of the cotton plant. Gossypol ($C_{30}H_{30}O_8$) is the most important seed pigment and forms close to half of the total intra-glandular pigments on the plant.

One of the most positive forms of host-plant resistance to insects comes from gossypols in cotton. Gossypol-free (glandless) varieties are attacked by insects more than varieties having gossypols. It is more important in seeds because gossypol is toxic to monogastric non-ruminants like swine and poultry. Gossypol-free varieties have been developed in many countries, but they are usually more susceptible to insects.

Gossypols, along with other terpenoids, occur in the sub-epidermal glands of the cotton plant. They are small spots on all parts of the plant. The boll wall is pitted on the outer side with gossypol glands, not very conspicuous in *G. hirsutum*, but black and more visible in other species. Among plant parts, gossypols are more prominent on the calyx of the flower.

On average, gossypols are around 1.5% of the total dry seed weight. Solvents used in soil extraction reduce the free gossypol in cake and meal. Gossypols can be deactivated with certain treatments like iron sulphate. Gossypol content is a varietal character controlled by two recessive genes. Gossypol-free varieties can be developed without any negative effect on yield.

A high gossypol content often means a high oil content. It is desirable to increase the gossypol content in the flower bud and other plant parts but reduce or eliminate it in the seed.

Gossypol-free cottonseed has many uses, including poultry rations and, particularly, in protein supplements as flour for baked goods in areas affected by food shortages.

COLORED COTTON

Cotton occurs naturally in four colors: white, brown, green and blue. White colored cotton ranges from creamy to shinning white. Brown occurs in various shades from light brown to dark brown and mahogany. Green occurs in shades from light green to green, but only very light blue is available. Some genotypes may show color fading with time and washings, while others may not. Green lint color fades in almost all the available green-color-producing genotypes. Brown color in some genotypes may intensify after many washings. Colored lint usually has poor quality: the fiber is weak, it has higher micronaire and it is shorter in length. The potential to improve these characteristics exists, but no better germplasm is available.

Lint color develops only after the boll opens and exposes lint to sunlight interaction. Lack of sufficient sunlight seriously affects the ability of a genotype to express its true color in the lint.

Depending upon varieties and intensity of sunlight, color development is completed in 6-10 days.

Colored cotton lint character is dominant over white lint in *G. hirsutum*, *G. arboreum* and *G. herbaceum*. The evolution of color is described as multiple allelalism and modifier gene action with intensifiers, suppressors and graded dominance of genes occupying linked loci. Colored cotton lint varieties have the same agronomic and pest control requirements as white cotton varieties.

Agronomy and Physiology

GROWTH AND FRUIT FORMATION

Photosynthesis is the plant's process of making food from water and carbon dioxide using light energy as the power source. The cotton plant synthesizes carbohydrates from carbon dioxide and water in the presence of light and chlorophyll (green pigment that serves as receptor of energy for photosynthesis). The process liberates oxygen in the air.

Photorespiration: The cotton plant releases a considerable amount of CO_2 during the photosynthesis process, a phenomenon known as *photorespiration.* Plants that photorespire belong to the group called C_3 plants. Photorespiration, which is virtually absent in C_4 plants, confers little or no benefit to the cotton plant. The rate of oxidation of photosynthetic products by photorespiration is about five times the rate of dark respiration. The respiration rate depends on temperature. High yielding cotton varieties have a lower rate of photorespiration than the less productive varieties. It is estimated that the cotton plant can lose about 30% of its assimilated carbon in photorespiration.

Cutout: In cotton, vegetative and reproductive phases can go side by side: a plant grows and forms bolls at the same time during most of its life. Cutout is the cessation of vegetative growth and fruit formation. *Cutout* is a stage determined by the cessation of growth due to the development of boll load sink and the resulting demand for available nutrients and photosynthate resources.

Cutout can be physiological, a stage beyond which no more productive bolls will be formed; or seasonal, a natural cutout; and premature. Premature cutout, which can happen due to stress, i.e., drought, pest damage, etc., can be corrected to put the plant into a productive stage again.

Why does cutout occur? A complete change from the vegetative to the reproductive phase occurs when the rate of dry matter accumulation equals the growth rate of the crop. At this stage, all photosynthates are channeled toward the existing bolls and new fruit forms are shed.

Cutout in cotton is always temporary. The cotton plant has the potential to get out of all kind of stress, terminate the cutout and start its regrowth and fruit formation again.

Following are some terms commonly used in discussions of cotton physiology.

Defoliation. Cotton leaves can be forced to drop off early by stimulating the formation of an abscission layer. Chemicals that have the ability to enhance the formation of an abscission layer and are used in the defoliation of cotton are called desiccants and defoliants. Cotton must be defoliated to be machine picked.

Effective fruiting period. The cotton plant forms a large number of flower buds but only some of them are turned into bolls. Under high temperatures, most early-formed flower buds are shed due to the inability of the pollen grains to germinate and fertilize ovules. But as soon as climatic conditions permit, flower buds become squares, flowers and ultimately bolls. The period beginning when the cotton plant starts forming squares and flowers that will become productive bolls through boll formation until cutout, is called the effective fruiting period.

Heat unit. Heat unit is a method of describing the growth and development of the crop utilizing temperature rather than age. Heat unit is also called degree-days. Heat units measure the amount of useful heat energy a cotton plant accumulates each day, each month, or for the season. The plant must accumulate a specific number of heat units to reach each development stage and to achieve developmental maturity.

Heat units or degree days can be calculated based on any lower temperature limit. The basic formula involves adding maximum and minimum temperatures, dividing by two and subtracting the threshold temperature (which could be any). In the case of DD60's, the lower limit is 60° F and is calculated as follows:

(Maximum temperature in F + Minimum temperature in F) /2-60

Lower temperatures produce less heat units and thus slow growth. Input use, pest pressure and climatic conditions have a strong influence on the crop development stage even if normal heat units have been achieved.

White flower. The freshly open flower in cotton is called white flower. Petals around the stigma and style, forming a candle-like structure, open up and expose anthers to direct sunlight. It happens on the day of anthesis/pollination. The flower is a white flower for less than 24 hours.

Pink flower. The white flower, fertilized or not, turns into a pink flower the next day.

Vigor index. Vigor index is the rate of growth in the main stem in relation to the number of nodes on the main stem. A higher vigor index means strong vegetative growth and a lower vigor index indicates reduced vegetative growth.

Plant mapping. One of the methods used to characterize structure and fruiting behavior of plants by recording branching and fruiting locations. Plant mapping permits monitoring of plant growth to see if it is behind or ahead of normal growth behavior. Plant maps are often used to determine nodes above white blooms, nodes above cracked bolls, fruit retention and to compartmentalize and compare fruit retention, on selected horizontal or vertical zones of the cotton plant.

Node. The node is a point where a branch or leaf is attached to the main stem. Nodes are usually counted on the main stem, though branches have similar attaching points. The attachment sites of the bolls on the sympodial branches are called positions. The first boll on the first node of a sympodial branch is usually called the "first position boll," on the second node, "second position boll," on the third node, "third position boll," and so on.

Soil Salinity. Cotton is classed as a salt tolerant crop. Varieties vary in their response to salinity, the accumulation of excessive quantities of soluble salts. A general threshold at which yield could start declining is 8ds/m, 25° C electrical conductivity.

Cotton is comparatively a salt tolerant crop, however, it is more sensitive to acidic soils than to alkaline soils. Cotton can be grown successfully on soils ranging from 6.0-8.0 pH values. However, the optimum pH for good cotton production on most soil types is close to neutral or 7. In salt-affected soils, seed germination is more of a problem than growth. Alkaline soil usually results in higher losses of nitrogen if a nitrogenous fertilizer is applied above surface and in ammonium form. Accordingly, an increase in soil acidity (lower pH) increases efficiency losses.

Rate of flowering. As the plant grows, flowers are formed in a vertical as well as a horizontal direction. A relationship can be established between the rate of boll formation in the vertical direction and horizontal direction. The relationship may vary depending upon varieties, but the rate of vertical boll formation is higher than horizontal boll formation.

Shedding. Square retention is the proportion of squares, usually expressed as a percentage, retained by the cotton plant. Boll position on the plant influences boll retention. First position bolls have the highest chances of being retained than later positions on the same branch.

Shedding of fruit forms, particularly flower buds, could occur due to many complex factors including meteorological, physiological, entomological, and nutritional. In spite of all favorable factors, some fruit forms, mostly flower buds, are bound to shed. The cotton plant is simply not able to retain all flower buds and convert them into bolls, and retains only as many bolls as it can afford to feed.

There are two main theories about shedding in cotton. One has to do with hormones, the other with carbohydrates. The balance between auxin and growth-retarding or anti-auxin hormones determines fruiting. As the boll load increases and a high number of buds and flowers are formed, production of anti-auxin hormones exceeds production of auxin. Fruit production decreases and shedding increases. The other theory is that as the number of bolls/fruit forms increases on the plant, supply of carbohydrates decreases and new fruit forms are shed.

Shedding of flower buds is a universal phenomenon in cotton. All varieties, without exception, shed large numbers of flower buds. Cotton has the highest compensatory ability among major field crops. If a flower bud, flower or a boll is shed, the cotton plant quickly tries to compensate that loss through production of more flower buds or even retaining buds that would otherwise have been shed. The maximum yield can be obtained when an optimum balance between the vegetative and fruit/bolls is maintained.

Average Growth and Fruiting of the Cotton Plant

STAGE	PERIOD	PLANT AGE
Planting to emergence	4 to 10 days	4-10 days
Emergence to first true leaf	8 days	12-18 days
Emergence to second true leaf	9 days	21-27 days
Second true leaf to pinhead square (seventh node)	18 to 21 days	39-48 days
Pinhead square to matchhead square	9 to 10 days	48-58 days
Matchhead square to first one-third grown square	3 to 6 days	51-64 days
First one-third grown square to first white bloom	12 to 16 days	63-80 days
First white bloom to first open bolls	40 to 60 days	103-140 days
Harvest bolls set on first four weeks of blooming	96%	91-128 days

Plant density. The number of plants required to get optimum yield depends on the plant type. The sympodial type (plant without monopodial/vegetative branches) requires a high number of plants for realizing a desired number of bolls per unit area, while the monopodial type has higher compensation ability and requires comparatively fewer plants for an optimum number of bolls per unit area.

Cotton is grown at various configurations, from ultra narrow row planting (six-inch row spacing) to skip-row planting where rows are already spaced at 0.5-1.0 meter. High plant density produces cotton plants taller than normal with short fruiting branches and high leaf area index, and contributes to boll rot. It also induces earliness as more bolls are positioned close to the main stem and because of a lack of monopodial branches. On the other hand, under low plant density conditions, the plant tries to compensate for the empty space, which helps fruiting branches to grow longer thus checking plant height.

Vegetative growth. A general term for undesirable cotton plant growth, often tall and rank, and usually bearing proportionally less fruit.

Rank growth. A term signifying tall, vegetative cotton growth; often a result of excessive nitrogen fertilizer, fertile soils and/or excessive moisture. Rank growth often renders cotton plants more attractive and susceptible to insects and boll rot, and more difficult to defoliate.

Terminal. The dominant, upper main stem of the cotton plant containing 3-4 expanding leaves and developing squares.

Square. The flower bud of a cotton plant when corolla (petals) is surrounded by bracts and still not visible.

Pinhead square/matchhead square. The earliest form of a square, when it becomes visible.

Fruiting position. A location on the main stem, vegetative branch or fruiting branch, where either a fruit is present or has aborted. The number of branches and bolls per plant is greatly influenced by plant spacing. Cold temperatures are the primary reason for a delay in the initiation of the first square and flower.

Root growth. Nitrogen application and water stimulate the growth of above-ground parts but suppress root growth. Defruiting has an increasing effect on the root system as well as carbohydrate content of the root. Optimum soil temperature for root growth is 28-35°C. Soil temperature below 10°C can have a chilling effect on the root tip up to five days after planting seed. Low soil temperature reduces root growth, water uptake ability and number of branches. Higher-than-normal soil temperature may temporarily increase branching. In general, most soils suitable for cotton production have 20% oxygen, 79% nitrogen and 0.15–0.65% carbon dioxide at a depth of six inches. Unlike most other crops, root growth in cotton is more sensitive to oxygen in the soil rather than to the CO_2 level. Standing water in cotton fields due to rain or over irrigation lowers the oxygen content in the soil thus suffocating the root system and causing the plants to wilt. If the effect is prolonged, plants die.

Defruiting. Increases plant height, internodal length and length of branches. High fruit retention retards growth.

Boll period. Boll period is the interval between anthesis and boll dehiscence/cracking. The boll period is a function of fiber elongation and the rate of cellulose deposition on the secondary wall. Both factors are strongly affected by temperature (first 15-21 days fiber elongation, and 21 days onwards fiber strength). Lower than normal temperature delays the initiation of fruiting forms. Low temperature is the strongest factor affecting fiber quality.

About 75-80% of the final boll dry weight is seedcotton; the remaining 20-25% of weight comes from tissues (burs).

More photosynthates per day per boll are required to shorten the period between flower opening and boll opening.

Bud formation starts 10-15 days prior to its visibility on the plant. The flower bud is more sensitive to water stress in the first week after its visibility. High temperatures three weeks prior to flower opening are more important under conditions where a lot of flowers are shed due to high temperatures. Developing bolls are the strongest sink for carbohydrates and plant nutrients.

Leaf area index. This is the ratio between the total leaf surface area of a plant and the surface area of ground covered by the plant. The primary effect of water stress is low photosynthetic activity due to low leaf area index.

Chlorosis. The yellowing or bleaching of normally green plant tissue usually caused by the loss of chlorophyll. The cotton plant being a deciduous perennial sheds its mature leaves naturally.

NUTRITION

Cotton needs carbon, hydrogen and oxygen, which are obtained from air and water. These nutrients are a must but their natural supply is enough and no additional artificial supply is required other than water. If the soil supply is not enough, they must be added to meet plant needs. Other nutrients needed for proper plant growth in cotton are calcium, manganese and sulfur. Among the micronutrients needed for cotton, boron can have some effect on plant health. Most nutrients other than macronutrients are usually available in most soils fit for cotton production. The quantity of fertilizer to be added depends upon soil structure and nutrient status, previous crops grown and target yield.

Nitrogen

Shortage and over supply of nitrogen have a significant effect on fiber quality. Shortage results in short and weak fiber, while excess supply can produce a longer fiber but weak and immature.

Nitrogen losses occur in soil due to volatilization and denitrification, as well as to leaching. Nitrogen can move in the soil quickly, and nitrogenous fertilizers have to be added more than once every year during the growing period. As the plant grows, the need for nitrogen increases and its availability in the soil has to be improved through synthetic fertilizers in successive doses. Soils low in organic matter are more susceptible to nitrogen deficiencies.

The first nitrogen deficiency symptoms are reduced growth, shorter height and fewer branches. Nitrogen deficiency also results in high fruit shedding and premature termination of fruit formation. A pale yellow color of leaves indicates nitrogen deficiency. Nitrogen deficiency at a later stage results in reddening of leaves. Yellow leaves are shed early. Excess supplies of nitrogen result in rank growth, delay in maturity and high pest attack.

Phosphorous

Phosphorous is used in growth regulation and has a minimum effect on fiber quality.

Alluvial soils are usually rich in phosphorous. In cotton, the availability of phosphorous is low when soils are over 7.5 pH (alkaline soils). Phosphorous moves very little in the soil and it is recommended to apply it before planting or at the time of planting and to work it well into the soil.

Phosphorous deficiency could have an adverse effect resulting in dark green leaves and stunted growth. Severe shortage may result in reddish purple leaves, reduced flowering and delayed maturity of set bolls.

Older leaves quickly translocate phosphorous to newly formed leaves, therefore, deficiency symptoms are more pronounced on older leaves.

Potassium

Potassium is required throughout plant growth but it is most needed at the time of boll maturation. Potassium is mostly needed for seed maturity but helps to maintain sufficient turgor pressure for fiber elongation, enzyme activation and pH balance in the cells–critical for plant health and disease suppression.

The leaves and stem in the cotton plant tend to maintain potassium concentration at high levels during the first 60 days after peak demand is reached, approximately three weeks after the first effective flower, lasting about six weeks. Bolls (in burs) carry more potassium than any other part of the plant.

Deficiency symptoms typically begin as a yellowish white molting in the area between leaf veins and at leaf margins. Leaf margins could be bronzed and curled downward as tissue breakdown continues. Bolls could remain unopened in case of severe potassium deficiency. Symptoms proceed from the bottom to the top of the plant.

Foliar application of potassium can have an effect on yield in high yielding areas if enough supply is not available from leaves. There is no difference between band application and broadcast application of potassium.

Boron

The most significant role of boron in the cotton plant's life, which is directly related to yield, is its effect on the pollen tube growth after pollination. Boron influences the conversion of nitrogen and carbohydrates into more complex substances such as proteins.

Boron deficiency is realized in acidic soils, while fine-textured calcareous soils are more deficient in zinc, iron and manganese.

Early stage boron deficiency may affect root tip elongation. At a later stage, the growing buds may show wilting or chlorosis of young leaves or buds.

Calcium

Calcium functions include strengthening of cell walls to prevent their collapse, enhancing cell division and plant growth, aiding in carbohydrate movement and balancing cell acidity.

Calcium deficiency in the soil affects the ability of the cotton plant to resist seedling diseases. A crop grown on calcium-deficient soil has a short internodal length, especially at shoot terminals, and weak stems.

Iron

Iron deficiency can be readily recognized from the inter-veinal chlorosis of young leaves. If the deficiency is not corrected, symptoms become more pronounced with each emerging leaf. Veins remain green, turning the rest of the leaf area chlorotic (lack of chlorophyll). Leaf margins may also show curling symptoms.

Magnesium

Magnesium is needed for the production of green pigments in chlorophyll, which is essential for photosynthesis.

Early magnesium deficiency symptoms resemble nitrogen deficiency symptoms, with dark green leaves developing red and purple spots. Symptoms can be confused with leaf aging. Symptoms may also appear on younger leaves as deficiency progresses.

Sulfur

Sulfur is essential for the production of three amino acids that are the building blocks in the synthesis of proteins. Organic matter is the primary storehouse of sulfur in the soil.

Sulfur deficiency symptoms on the plant can be seen in the form of irregular thickening of young petioles. Pale green to yellow leaves appear in the upper part of the plant. The petiole effect is more consistent and the petiole and stem may be splitting in case of severe sulfur deficiency. Internodal length is also short.

Zinc

Zinc deficiency symptoms on the plant are in the form of small but thickened, brittle leaves with interveinal chlorotic symptoms on leaves. The plant becomes bushy and green leaves are shed.

Other nutrients such as copper, manganese and molybdenum also show deficiency effects on the plant, mostly in the form of chlorosis and small yellowing/dead tissues on older leaves.

WEED CONTROL

Weeds are pests and take a heavy toll on the plant by sharing inputs applied for cotton growth. Weeds also harbor insects and pathogens. Weeds affect cotton yield and quality. A reduction in yield is mainly due to a smaller number of bolls per plant. The weed population is not static but changes in response to growing conditions, control procedures and cropping sequence. For example, narrow row spacing reduces weeds. Weeds can be removed manually, mechanically, biologically and chemically.

Hand hoeing and weeding is primarily done in small scale farming systems. Tilling is another method of weed control. In addition to removing weeds, tillage modifies the physical condition of the soil such as bulk density, aggregation size, penetrability, etc. These altered physical properties affect biological and chemical processes occurring in the soil that consequently affect growth and yield.

Crop rotation and fallow period have also been successfully used in controlling weeds, but with the high demand to increase cropping intensity their role has diminished.

The biological control of weeds that involves the use of microorganisms was recognized in 1940; however, biological control never became a major means to control weeds in cotton.

Herbicides are chemicals used to control weeds. Herbicides can be divided into pre-emergence and post-emergence, selective and non-selective, and contact and systemic. The mechanism used by a herbicide to kill a weed is called mode of action. The mode of action of herbicides on the cotton plant varies greatly depending upon the chemistry of each product. The primary modes of action that are used in cotton are growth regulators or hormone herbicides, photosynthesis inhibitors, pigment inhibitors, seedling growth inhibitors, contact, lipid synthesis inhibitors and amino acid synthesis inhibitors. A herbicide can be applied to all the field, a portion of the field or only directed to specific weeds. Lay-by application is a final, typically post-directed herbicide application designed to eliminate or suppress weeds by directly hitting them in spots.

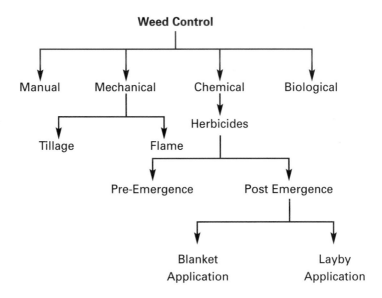

Very little work has been done on the biology of weeds in cotton. Some commonly occurring weeds in cotton are:

Technical Name	Common Name
Abutilon theophrasti	Velvetleaf
Amaranthus spp.	Pigweed
Celosia argentea	Cock's comb
Chenopodium album	Lambsquarters
Convolvulus arvensis	Field bindweed
Cynodon dactylon	Bermuda grass
Cyperus esculentus	Yellow nutsedge
Cyprus rotundus	Purple nutsedge
Dactyloctenium aegyptium	Crowfoot grass
Datura spp.	Datura
Digitaria sanguinalis	Crabgrass
Echinochloa colonum	Junglegrass
Echinochloa crus-galli	Barnyard grass
Hemrizonia pungens	Spike weed
Ipomoea spp.	Morning-glories
Panicum colannum	Rabbit grass
Portulaca oleracea	Purselane
Setaria spp.	Foxtail
Sorghum halepense	Johnson grass
Solanum nigrum	Black nightshade
Solanum sarrachoides	Night shade
Vicia sativa	Common vetch
Trianthema monogyna	Santhi
Xanthium spp.	Cocklebur

4 Diseases

Diseases are difficult to detect. The most popular control methods are cultural and host plant resistance. Biological and chemical controls are expensive and not popular in cotton. The only chemical control method against diseases is seed treatment with fungicides, bactericides and other chemicals. The most important diseases are caused by pathogens in the soil. Although soil treatment can be a control measure in addition to cultural, biological, host plant resistance and chemical control, it is impractical. Fungi, bacteria, nematodes and viruses can cause seedling diseases in cotton.

Root Rot

Causal organism

Phymatotrichum omnivorum (originally called Ozonium auricomum, Ozonium omnivorum and finally in 1916 named Phymatotrichum omnivorum) is the most common cause of root rot in cotton.

Symptoms

Include bronzing or slight yellowing of leaves, discoloration of xylem and increases in leaf temperature. Dry leaves do not shed. The main root is affected before the lateral roots.

Control

Genetic resistance is the most effective means of control, but currently no varieties with resistant genes are available.

Seeding Diseases

Causal organism

Many species of *Pythium* are known to cause seed rot and are the most common cause of pre-emergence damping-off in cotton, but *Pythium ultimatum* is more prevalent. At an early stage, damping-off symptoms are similar in both fungi. Other fungi, like *Rhizoctonia solani*, *Thielaviopsis* and various *Fusarium* spp. can also cause damping off. *Xanthomonas campestris* pv. *malvacearum*, the bacterium that causes bacterial blight, can also cause seedling disease. *Rhizoctonia solani* is the most common cause of post-emergence damping off and it can cause seed decay, pre-emergence damping off if seed is planted in cool soil, lesions on the hypocotyls and root rot.

Symptoms

Seedling disease can manifest in different stages from pre-emergence seed rot to post-emergence damping off. In damping-off caused by *Pythium*, seedlings fail to emerge out of the soil. *Pythium* spp. and *Rhizoctonia solani* can cause post-emergence damping-off, and symptoms in both cases are similar. Certain *Pythium* spp. is more damaging at low temperatures and high humidity. Low soil temperatures encourage infection by R. solani. If only minor roots are affected by *Rhizoctonia solani*, only stunting- like symptoms may be seen. At seedling stage, *Colletotrichum* sp. has the potential to cause post-emergence damping-off, but it can also affect the plant at any stage of development. The fungus is carried on the seed and within the seed. Fungicides are effective to control the disease, but in order to avoid infection from within the seed, it is recommended to use planting seed from areas not affected by *Colletotrichum*.

Control

It is very difficult to eradicated soil-borne fungi like *Pythium* spp., *Rhizoctonia solani*, *Thielaviopsis basicola*, *Colletotrichum* sp. and *Fusarium* spp. once it has established in the soil. Fungicides, applied as seed dressing or in the soil at planting, are the only solution to control these pathogens. Damping-off caused by *Rhizoctonia solani* can be controlled effectively with fungicides applied as seed treatment or by in-furrow application at planting. A strong genetic resistance source is not available. Any factor that encourages rapid emergence and growth of the seedling is helpful in avoiding the disease.

Black Root Rot

Causal organism

The fungus *Thielaviopsis basicola* is responsible for the disease commonly known as black root rot.

Symptoms

The disease is more severe in clay soils and causes damage only within a temperature range of 20°-36°C.

Control

Fungicide use is usually not economical. It is recommended to delay planting until soil temperature is above 20°C. Avoiding over-irrigation also helps to contain the disease.

Charcoal Rot

Causal organism

Charcoal rot is caused by *Macrophomina phaseolina*, an equivalent of *Sclerotium bataticola* and *Rhizoctonia bataticola*. The soil-borne fungus can survive in infected seeds.

Symptoms

The main root and stem are often affected, but leaf spots can also occur. No genetic resistance to the disease has been found, but cultural practices, like interculturing, help to control it.

Verticillium Wilt

Causal organism

Verticillium wilt, caused by the fungus *Verticillium dahliae* is the most prevalent disease in cotton, particularly in irrigated areas. Due to the controversy over the taxonomy of *Verticillium* species, prior to 1970 verticillium wilt was believed to be caused by *Verticillium albo-atrum*. Now it is accepted that *Verticillium dahliae* is responsible.

Symptoms

Disease symptoms vary according to the growth stage of the crop. The crop is usually attacked in the beginning or at the end of the season, when the mean soil temperature is below 29°C. Plants may not be killed, but leaf shedding, shedding of small bolls, and premature opening of large bolls are common symptoms if the crop is affected at boll formation or maturation stage. Early season symptoms appear in the form of mosaic patterns on leaves of infected plants indicating that the attack is serious. Yellow mosaic symptoms first appear on older leaves and then higher up in the plant. Symptoms first appear on leaf margins and then on the area between veins. A cross-section of the stem of an infected plant will reveal tan to brown flecks in the main stem. These flecks are due to the formation of melanized (darkened) products in the infected xylem vessels and surrounding cells.

Control

Cotton monoculture has favored the establishment of the pathogen in the soil. Plants with excessive vegetative growth/succulent plants are more easily affected by verticillium wilt. It is important to maintain a balanced nutrient status in the soil and in the plant. Under certain nutrient deficiency conditions, like potassium deficiency, plants will be more susceptible to infection. Plants subjected to excessive levels of nitrogen fertilizers will also be more susceptible to infection. The application of gin waste to fields from crops infected with verticillium wilt will increase the levels of the pathogen in the soil. Variety monoculture or the cultivation of susceptible varieties will also increase levels of the pathogen in the soil. *Verticillium dahliae* is very sensitive to soil temperature. Areas and regions where soil temperature remains above 30°C for a long period of time are not likely to be affected by verticillium wilt. A strong genetic control that has been extensively utilized in the world is available, but all varieties show good resistance to verticillium wilt in areas where the mean temperature is above 30°C at early stages of plant growth. Variable responses by different varieties indicate that multigenes expressing an additive effect control resistance to verticillium wilt. The basis of resistance seems to be a physical barrier in the xylem prohibiting the pathogen to spread in the plant. Chemical and biological control are not feasible, as the pathogen cannot be eliminated from the soil. A combination of in-built resistance—with agronomic measures including crop rotation, elimination of crop residue and acid delinting of planting seed—is the best defense.

Fusarium Wilt

Causal organism

Fusarium wilt, caused by *Fusarium oxysporum*, was first discovered in 1892 in the USA. The first reports about this disease outside the U.S. came from Egypt in 1902. Fusarium wilt is a disease of warm climate. The most desirable temperature is 30-32°C and the most desirable soil moisture level is 85% saturation. Nitrogenous fertilizers enhance the chances of disease incidence, while potassium fertilizers have a suppressing effect. Phosphorous has no effect. The presence of root-knot nematodes in the soil increases the chances of wilt incidence.

Symptoms

Disease symptoms are very similar to those of verticillium wilt, though both wilts rarely occur in the same field because of their response to temperature. Vascular browning of stem tissues inside can be seen by a naked eye with a cross-section of the stem. Leaf margins show a lack of chlorophyll and ultimately the entire leaf turns yellow. Healthy, normal looking leaves and leaves showing disease symptoms appear on the same plant, particularly if the disease appears at a late stage of plant growth. Infected parts fail to transmit water to lateral parts of the plant, resulting in water deficit and wilting. Along with yellowing leaves, plants show stunted growth. Plants may survive and recover if all leaves have not been shed before conditions unfavorable for the disease appear. Fusarium wilt is less sensitive to low temperatures compared to verticillium wilt. Fusarium wilt symptoms can appear at any stage of crop development depending upon the occurrence intensity of the inoculums in the soil and host plant resistance to the pathogen. In severe attacks the plant may be killed, particularly at an early stage. Yield loss depends on when and how much of the plant is affected. The fusarium wilt pathogen is carried in the seed and usually, most seeds from a plant showing vascular browning are infected. Planting seed carrying the pathogen can infest soils previously free of the pathogen. The pathogen is also carried physically to healthy fields by various means, like farm implements.

Control

Once the pathogen is established in the soil, it is not easy to eradicate. The pathogen can survive in the soil for many years and infect when suitable conditions and crops become available. Soil fumigation can help, but usually it is not recommended for other than high value cash crops because it is not economical.

Rotation is the best and least expensive control method when integrated with resistant varieties. Disease resistance has a polygenic control. The best screening of resistant genetic material can be carried out on established hot spots.

Some Distinguishing Features of Fusarium and Verticillium Wilts*	
Fusarium Wilt	Verticillium Wilt
Favored by high temperatures, i.e., daily mean above 23°C.	Favored by low soil temperatures i.e., daily mean below 23°C.
Most favorable soil type varies with race of the pathogen. Race 1 is favored by light soils of neutral to acid pH.	Favored by heavy soils of neutral to alkaline pH.
Symptoms can appear at any stage of crop development, but usually begin 4 weeks after planting. Plants become more susceptible at flowering.	The disease can first be found in the crop about 6 weeks after planting but its incidence usually increases toward the end of the cropping season as temperatures begin to fall.
Symptoms are similar to verticillium wilt, but chlorosis tends to be in patches without reddening. Part of the leaf may be chlorotic and flaccid with the rest appearing healthy. In advanced stages of the disease, extensive defoliation occurs. Vascular browning is pronounced, extending into the outer stele tissues.	Symptoms begin with marginal chlorosis and, sometimes, reddening of the leaf. Chlorosis develops between the veins. The rate and extent of defoliation is dependent on the strain of the pathogen. Vascular browning is less pronounced than with fusarium wilt.

*From: Hillocks, R. J.1992. Cotton Diseases, C.A.B. International, Walingford, Oxon OX10 8DE, UK.

Bacterial Blight

Causal organism

The disease is caused by the bacterium, *Xanthomonas campestris pv. malvacearum*. Bacterial blight is the collective name for the different manifestations of this disease (seedling blight, black arm, stem canker, angular leaf spot and bacterial boll rot). Countries with hot and dry weather during the cotton-growing season have less chance of getting bacterial blight.

Symptoms

Typical leaf symptoms are the appearance of small, water-soaked angular lesions on leaves. The lesions can occur along the main margins of leaves, causing veinal necrosis. Older leaves may show chlorotic patches, usually smaller than the angular spots. Under hot, humid conditions, the disease can also infect stems and bolls, causing severe boll rot. Spots on bolls appear as oily, round lesions. The bacteria can spread internally to the seed and survive for many years. Contaminated seed and crop residue are the primary source of infection. Irrigation water run-off can spread the disease from one field to another if the infection source is plant debris. Plants can be infected at any stage but only under wet conditions. The pathogen has many races, and more than one race has been found in many countries. Sucking insects can also transmit the disease.

Control

Seed treatment is a very effective control measure. *G. barbadense* has the highest susceptibility to the disease. *G. hirsutum* has a wide variety of resistant genes. Varieties resistant to more than one race at the same time have been developed successfully in many countries

Nematodes

Causal organism

More than thirty genera and 128 species of nematodes are reported as pests of cotton but not all of them are harmful. Some species of nematodes may not be parasitic. On a worldwide basis, five species of some concern to cotton growers are:

Meloidogyne incognita	Root knot nematode
Rotylenchulus reniformis	Reniform nematode
Hoplolaimus galeatus	Lace nematode
Hoplolaimus columbus	Lace nematode
Belonolaimus spp.	Sting nematode

The root knot nematode *Meloidogyne incognita* (race 3 and 4) is the most commonly occurring nematode in cotton fields. None of the five common species of nematodes can be seen with the naked eye, but there are other species that can. Nematodes, like many other pathogens, occur in hot spots that serve as infection sites for further multiplication. Nematodes are assessed in terms of weight per sample weight of soil.

Symptoms

Nematodes cause physical injuries to the root, which affect growth in two ways:

Plants form shorter and fewer internodes, thus affecting plant height. Symptoms usually appear at the time of first bloom. If the infection is severe, leaves may turn pale yellow, similar to symptoms of nitrogen deficiency and lack of photosynthetic activity. Leaves may ultimately fall.

Some parasitic nematodes may form galls on roots and block the flow of nutrients from the root to the upper part of the plant. Plants may show wilting symptoms and green leaves may be shed in case of severe infection. The symptoms are usually observed with infection by root knot.

Nematodes can affect yield as well as quality depending on the stage of the crop when the disease is spread. *M. incognita* is more damaging to cotton in sandy soils compared to clay and loam soils. Nematodes can occur in association with fusarium wilt. High temperature is injurious to nematodes. In areas where the atmospheric temperature goes above 40°C, nematodes are usually not a problem. Nematodes can multiply at a faster rate in soils that have a moisture content of 40-60% of field capacity. These conditions cannot exist in soils where the atmospheric temperature exceeds 40°C.

Control

Chemical control is effective. Fumigants are more effective than non-fumigants and may even have residual control on the following cotton crop. Varietal tolerance is also available as a component of control measures. However, an integrated management strategy, including chemical control, tolerant varieties, cultural control measures, rotation of non-host crop, sub-soiling and soil solarization are more suitable for a long term solution.

Virus Diseases

Causal organisms

The following viruses, which are the pathogens, cause the most important diseases in cotton:

Cotton leaf curl
Cotton leaf crumple
African cotton mosaic
Cotton blue disease

Disease names are based on symptoms caused by the viruses.

Symptoms

In addition to symptoms evident from the name, the most common symptoms of virus disease are stunted growth, reduced flowering and fruit shedding. Yield loss can vary greatly—if the plant is affected at an early stage, the loss in yield can be huge. Virus diseases are usually transmitted by insect vectors, and in the case of cotton, it is mostly transmitted by whiteflies or aphids. One of the limitations of acquiring more knowledge about virus-causing diseases in cotton has been the inability to extract and transmit viruses for experimental purposes.

Control

There is no chemical control of viruses. The most effective control lies in the control of the vector. The vector population has to be reduced to levels lower than those of the vector's threshold as a direct pest on cotton. Other methods of control are the elimination of alternate host plants and seasonal weeds, and the destruction of infected crop residue. Resistant varieties have been developed.

Boll Rot

Causal organisms

A number of fungi and bacteria can enter the boll and cause boll rot, resulting in total yield loss. Some causal organisms are *Diplodia gossypina*, *Colletotrichum* spp. and *Fusarium* spp. Boll rot can also occur due to *Rhizoctonia* spp. and *Alternaria* spp. The bacterial blight pathogen *Xanthomonas campestris pv. malvacearum* can also cause boll rot.

Symptoms

Bolls of all ages can be affected, but older bolls at the bottom of the plant are more likely to get infected. Wet weather and dense crop canopy aggravate boll rot.

Control

Chemical control is usually ineffective. Bract (outside covering of a boll) size is another criteria to determine the extent of the disease. Frego bract (smaller in size) varieties are less susceptible to boll rot disease.

Cercospora Leaf Spot

Causal organism

Cercospora leaf spot usually occurs along with alternaria leaf spot and is caused by *Cercospora gossypina*.

Symptoms

Disease symptoms show as irregular brown lesions surrounded by chlorotic tissues. Cercospora leaf spot does not cause any significant loss in yield. Infested undecomposed leaves are a major source of the disease. High humidity favors infection and spread. Older leaves at the bottom of the plant get infected first.

Control

Foliar application of fungicides has only limited success.

False Mildew

Causal organism

False mildew is caused by the fungus *Ramularia areola*. In some stages of its life cycle, the fungus is known as *Cercosporella gossypii*.

Symptoms

False mildew is found more frequently in dense crops. Wet weather favors the spread of the disease. The disease appears late in the season when it cannot cause any significant loss in yield. The fungus survives on infested leaf debris between seasons.

Control

Fungicides are effective. Some varietal resistance exists.

Alternaria Leaf Spot

Causal organism

Alternaria leaf spot is caused by a variety of fungi of which the most important are *Alternaria alternata* and *Alternaria macrospora*. These two fungi often occur together in the same plant.

Symptoms

G. barbadense is more susceptible compared to other cultivated species. Some *G. hirsutum* varieties have good resistance to these pathogens. *G. herbaceum* is equally affected. Leaves formed in the early growth stages of cotton have higher chances of becoming infected compared to leaves that develop later in the season.

Control

Some fungicides are effective against both pathogens. The pathogens are carried in the seed, but undecomposed leaves are also a major source of disease infestation. High humidity encourages the spread of the disease, but high temperatures, close to 40°C, are not favorable for the disease. Resistant varieties are available.

Rhizoctonia Leaf Spot

Causal organism

It is caused by *Rhizoctonia solani*, a seedling pathogen occasionally affecting foliage in plants with rank growth.

Lint Contamination

Causal organism

Many fungi that cause boll rot are also capable of contaminating lint. Fungi can easily move from boll surface to lint and cause discoloration which affects lint value.

Symptoms

Humid weather conditions for many days and shade encourage boll rot and also lint contamination. Cotton in bolls located at the bottom of dense, leafy plants is more prone to such infection. Heavy whitefly infestation that contaminates cotton through sugary secretions stimulates fungi growth on the honeydew-contaminated cotton. Open bolls not picked for many days and exposed to natural dew are more susceptible to developing fungi.

Some Other Diseases

Disease	Causal Organism
Anthracnose	*Colletotrichum gossypii*
Phomosis leaf spot	*Phomosis malvacearum*
Phoma leaf spot	*Phoma exigua*
Cochliobolus leaf spot	*Cochliobolus spicifer*
Myrothecium leaf spot	*Myrothecium roridum*
Powdery mildew	*Leveillula taurica*
	Salmonia malachrae

5 *Insects*

The cotton plant is naturally vulnerable to a variety of insect pests. A broad-based approach to control insect pests and save the crop from losses must be developed based on fundamental knowledge about the pest, crop and growing conditions. Proper crop protection measures not only save the crop from losses but also improve the effectiveness of inputs applied to grow cotton and improve fiber quality.

INSECT IDENTIFICATION, BIOLOGY AND NATURE OF DAMAGE

The main aspects of pests discussed here are identification, life history, and nature of damage to cotton. Like other organisms, the life cycle of various insects is dependent on many factors. The average life of developmental stages of pests given here is for an optimum situation to determine the feeding period on the cotton plant.

Sucking Insects

Acrosternum hilare and Chlorochroa ligata (Stink bugs)

Numerous species of stink bugs, family Pentatomidae, can be found on cotton but many are predacious (living on prey) and only a few cause economic losses in cotton. *Acrosternum hilare* and *Chlorochroa ligata* are important in most cotton producing countries. Both species are shield-shaped and flat. The green stink bug *Acrosternum hilare* is bright green, while the conchuela stink bug *Chlorochroa ligata* is dark brown to black with a red border and a red spot on the tip of the abdomen. The life cycle of both species is similar and completed in about 40-50 days.

Stink bug eggs have a distinctive barrel shape and are usually laid in clusters on stems and leaves. These egg masses resemble many barrels lined up in rows. Eggs hatch in a few days.

Stink bugs have five nymphal instars (development stages) and the average length of time for each instar is 4, 6, 8, 10 and 12 days for the first through fifth instars, respectively. Nymphs resemble adults with developing wing pads becoming visible in the fourth instar.

Stink bugs feed by inserting their long, piercing mouthparts and sucking out plant sap, thereby injuring squares and bolls. Infected bolls may fall off the plant, or if they remain on the plant, they have hardened dry locks with unspinable, stained, low quality lint.

Amrasca devastans (Jassid)

Amrasca devastans, previously known as *Amrasca biguttula* or *Empoasca devastans*, is called Indian cotton jassid or potato leaf hopper. *Amrasca devastans* is more common than other types of jassids. *Jacobiasca lybica* and *Jacobiasca facialis* occur in Africa.

Amrasca devastans eggs embedded in the midrib or larger veins on either side of the leaf or petioles hatch in about 5-10 days depending on temperature.

Nymphs are pale yellowish green in color and pass through five stages that last about 14 days. During this period, their length increases from 0.5 mm to 2.5 mm., and they are confined to the undersurface of the leaf during day time.

Adults are about 2.5 mm long and greenish in color. They easily fly when the plant is shaken. Both, adults and nymphs, feed on small veins on the underside of the cotton leaf.

Jassids inject a toxin into cotton leaves that impairs photosynthesis and causes leaf edges to curl downward in young leaves. This interferes with translocation in the phloem and leaves to the edges. Leaves first turn pale green, then yellow and finally red. Jassid prefers smooth leaf varieties, and in severe attacks plant growth is stunted. Nymph cause more damage than adults. The total life cycle of the pest is 3-4 weeks.

Biological control does not control jassid effectively. Leaf hairiness provides a strong control and has been effectively used by breeders in many countries. Longer and dense leaf hairs limit oviposition (egg laying).

Aphis gossypii (Aphid)

Aphid Aphis gossypii, also called cotton or melon aphid, belongs to Homopetara and has a cosmopolitan distribution. Another aphid species of the same importance on cotton is *Acyrthosiphon gossypii*.

Aphis gossypii is polyphagous (eating a variety of food) and varies in color from dirty green or grayish green to dark green and sometimes even blackish brown. Cotton is a favorite host but aphids commonly occur on many other plants like okra, cucurbits, and leguminous crops.

The immature or nymphal stage looks like the adult stage and the only difference is smaller size during the immature stage. Greenish winged females reach a cotton field and establish their colonies. The females are 1.5-2.0 mm long with red eyes and black spines at the posterior end. Females give birth to yellow nymphs that become globular wingless females with prominent siphunculi. These nymphs continue to breed without mating and laying eggs (parthenogenesis).

Aphids are always found on the undersurface of cotton leaves and reproduce rapidly. One female can produce about 80 young females parthenogenitically, which can mature within 8-10 days. Under optimum conditions, about 50 generations can occur each season, one each 5-7 days.

Aphids can damage cotton in two ways: heavy attacks can interfere with transpiration and photosynthesis, and they can spoil lint quality by discoloration and honeydew secretion. In addition to the effects as a sucking pest on cotton, aphids also serve as a vector for transmission of diseases and stickiness. Generally, aphids attach to younger leaves, which in the case of a heavy attack may form cup shaped structures downward. Aphids are rarely found on squares and flowers and do not induce shedding.

Natural biological control through parasitoids and predators exist in most countries. The best cultural control is to avoid planting alternate hosts near cotton. Chemical control is very effective, and this pest is often controlled when other sucking pests are also treated with insecticides.

Bemisia tabaci and Bemisia argentifolii (Whiteflies)

Two species of whitefly damage cotton, i.e., *Bemisia tabaci* (sweet potato whitefly, cotton whitefly, tobacco whitefly, cassava whitefly, biotype A) and *Bemisia argentifolii* (silver leaf whitefly, poinsettia strain, biotype B).

Bemisia tabaci is more prevalent than the other species. Many species of whitefly are similar in morphology and difficult to differentiate. Whitefly species have a high elasticity in their morphology which changes significantly with changes in the host range. The fourth instar "pupal case" is commonly used for identification purposes. *Bemisia tabaci* and *Bemisia argentifolii* differ in esterase banding pattern, DNA composition, pathogen transmission characteristics, honeydew production and egg laying.

The whitefly life cycle is comprised of six stages: egg, three nymphal stages and a fourth pupal stage with two often recognized entities, and adult.

Whitefly is mobile on the plant and the female prefers younger leaves for oviposition. Most of the eggs are laid on fully opened leaves, from the 5th to the 8th nodes from the terminal. The number of eggs per life span is 100-300 per female. Eggs require about 5 to 6 days to hatch.

Both species of whitefly damaging cotton are polyphagous. *G. hirsutum* varieties are more susceptible to whitefly. Whitefly prefers hairy leaves, thus less hairiness and okra leaf type (less hairy leaf area) are not favorable for multiplication.

Adults are first whitish yellow but after a few hours of emergence change to completely white. On average, male and females measure 1.2 to 1.4 mm in length, respectively. They are bright white in color and actively fly when disturbed. Newly emerged adults start mating from a few hours to 1-2 days after emergence. Pairing starts in the morning and the copulation period is usually less than four minutes. At 30°C the life cycle is completed in 17 days but may extend to six weeks, longer in winter and shorter in summer.

Whitefly damages the plant during two stages of life. The adult stage is the most damaging, but nymphs also feed on cotton leaves and they inject saliva. Adults and all nymphal instars excrete copious amounts of honeydew on leaves and open bolls, which create stickiness and provide a medium for the growth of fungi. Losses are qualitative and quantitative in nature. Whitefly is also known as a vector for the transmission of about twenty viral diseases, the most important being leaf crumple and leaf curl diseases.

Chemical control of whitefly is effective with insecticides that are toxic to both adults and nymphs. Whitefly is known to develop resistance quickly.

Dysdercus spp. (Cotton stainers)

Many species of *Dysdercus* are known to damage cotton, particularly long staple. *Dysdercus cingulatus* is one of the species that also affects upland cotton.

Adult stainers are 15 mm long and 4.5 mm wide and have piercing, sucking mouthparts. The head and pronotum (dorsal body plate of the first section of the thorax, which are frequently enlarged and prolonged in many insects) are bright red or reddish orange; the body is dark brown crossed with pale yellow lines. The antennae are about 10 mm long. There are species differences, and the color can also be yellow and orange. The variability is high and adults can survive just on water for many weeks after the cotton season.

Eggs are smooth and initially creamy white, changing to orange as the embryo develops. Eggs are laid in the soil or under debris and hatch in 5 to 13 days depending on temperature, humidity and species.

Immature stainers undergo five instars with a total life span ranging from 21 to 35 days. The second and third instars feed gregariously on bolls. Rudimentary wings appear at the third instar. The fourth instar is larger and square-shaped.

Stainers damage developing bolls by puncturing seeds and causing plant sap to ooze. Plant sap stains lint with an indelible yellow color. Stains also affect plant mass and seed oil content and hence cotton's total value.

Stainers are susceptible to a variety of insecticides.

Lygus lineolaris (Tarnished plant bug)

Adult tarnished plant bugs are 8-10 mm long and 4-5 mm in width, 2-3 times larger than cotton fleahoppers. They are generally brown with splotches of white, yellow and reddish-brown. Along the side of the body on the hind part of the wing there is a clear, black-tipped triangle.

Tarnished plant bug eggs are white and slightly curved and hatch in 8-10 days. Nymphs are pale green and have an orange spot in the middle of the abdomen. After feeding begins, nymphs become darker green and black spots become evident on the body segment (thorax) between the head and abdomen. The nymphal stage is completed in 10-13 days. In the summer, the pest completes its life cycle in 20-30 days.

Tarnished plant bugs suck sap from the plant, particularly from flower buds and small bolls, which may be shed. Injured bolls that are not shed have damaged lint. Damage may also range from ragging brown discolored tissue, premature drop of buds, flowers, and fruit; an increased number of vegetative branches; multiple crowns; elongation of internodes; split stem lesions; swollen nodes; and leaf crinkling.

The removal of preferred host plants helps to reduce damage. Several insecticides are available to control populations of *Lygus lineolaris*. Several parasitoids are known to control the bug, but some egg parasites are thought to be relatively more important.

Tetranychus spp. (Mites)

Mites have eight legs in the adult stage instead of six which is a characteristic of insects. Adults, located on the underside of the leaf, are so small that they can barely be seen with the naked eye. A number of mite species damage cotton, and they all have almost the same biology. They are usually red, but may be green, orange or straw colored. *Tetranychus aurticae* (Spider mite) is one of the most common species. The spider mite is brick red or sometimes reddish yellow.

Spider mites pass through egg, nymphal and adult stages. They overwinter in the adult female stage in protected vegetation near the target crop fields. Females lay egg masses on the underside of leaves. Eggs hatch in 2-5 days.

Nymphs feed for 2-3 days before molting (shedding their skin) and developing into eight-legged nymphal mites. Nymphs feed for 5-6 days before turning into eight-legged adults. Adults feed for 1 to 3 days before laying eggs. The entire life cycle is completed within a week under optimum conditions. Spider mites have an unusual life history as the fertilized eggs produce females and unfertilized eggs hatch into males.

In many countries, the misuse of insecticides, particularly early spraying of pyrethroids, is tied to mite attacks at late stages of the crop. Mites appear in patches in the field and their attack is more severe under dry conditions at or close to the crop maturity stage. Nymphs and adults suck sap from leaves and affect photosynthetic activities. The upper surface of leaves becomes pale green and reddish brown initially, and may turn bronze under heavy attack. Leaves may turn up if the pest population is high.

Spraying only the affected patches in a field can control mites. Pyrethroids should not be used extensively at an early stage. Many acaricides can effectively control mites.

Thrips spp. and Frankliniella spp. (Thrips)

Thrips are about 2 mm in length and are usually yellowish-brown, black, tan or orange in color. Thrips can reproduce without mating. Mated females produce both males and females while unmated females produce only males. Males are wingless, while females have long narrow strap-like wings, as in *Thrips tabaci*, a more common species than others.

Eggs are laid inside plant tissue and hatch within 5 days. The three nymphal stages are completed in about 6 days, and nymphs do not have wings. Including 4-6 days of a pupal stage, egg-to-adult development is completed in about 16 days. The average life span of a mated female is 8-20 days.

Thrips nymphs and adults suck sap from the lower surface of leaves and damage terminal buds. Their feeding ruptures cells which causes stunted plants and wrinkled leaves that curl upward. Severely affected leaves get silvery and give a shining look on the underside. The photosynthetic process is affected, and the plant fails to grow normally. Weaker plants shed younger buds and give smaller bolls, which results in lower yields.

Attacks quickly spread on plants and damage can be noticed at some distance without entering the field. The pest population is usually evenly spread throughout the field.

Feeders

Agrotis ipsilon (Black cutworms)

The commonly occurring cutworms in cotton are *Feltia subterranean* (granulate cutworm), *Peridroma saucia* (variegated cutworm) and *Agrotis ipsilon* (black cutworm). However, black cutworm is more prevalent and found in most countries where cutworms are a serious pest on cotton.

The adult cutworm is a brown to grey moth with a greasy look. Wings are 25-40 mm in span. Several species of Agrotis infest cotton. Cutworm females lay their eggs generally on grass or in the soil in low spots of the field. Eggs normally hatch in 3 to 5 days. A female lays about 500 eggs.

Larvae are often shiny or glossy and always have four pairs of pseudo legs near the posterior end and three pairs on the anterior end. Cutworm larvae characteristically curl up into a C-shaped loop and come back to the same position readily if disturbed. Cutworm larvae usually cut off the apical portion of seedlings (young shoots). Cutworms also feed on leaves when the plant is fully grown. The average time of larval feeding is 2 to 3 weeks. Their life cycle is completed in about 30 days. Damage is more visible at field edges.

Most cutworms overwinter in the larval stage, but others overwinter as pupae or adults. The pupal stage remains in the soil for 7 to 8 days.

If damage occurs at an early stage, the entire plant maybe lost. Pre-planting herbicides and tillage help to control cutworms.

Alabama argillacea (Cotton leafworm)

The adult cotton leafworm is a tan to brownish moth with a few darker, wavy, transverse bars on the forewings. There may be an olive gray or purple tinge to the coloration. Distinctive characteristics of cotton leafworm adults are undulating reddish lines across the front wings, and the oval dark spot near the center of each forewing. The wing span is close to 40 mm.

The female starts ovipositing within 2 to 4 days after emergence. A single female may lay up to 600 eggs scattered singly on the underside of leaves. Eggs hatch within 3 days.

Cotton leafworm larvae are smooth, light to dark yellowish-green and marked with three narrow white stripes dorsally and one laterally. The larvae have four black dots that form a square on the top of each body segment, similar to a double-four domino. Each black spot has a black spine, and the center is surrounded with a white ring. The young larvae begin feeding on cotton leaves and pass through six to seven instars to complete larval development. The average life of a worm under optimum conditions is about 15 days. Fully fed larvae are 40 mm long and change to a brighter color as they mature.

Slender green larvae are semi loopers and have two broad black velvet stripes separated by a thin white line and more black and white stripes towards the sides. Black spots are scattered throughout the body.

Full grown larvae of cotton leafworms web up with a boll or roll of leaves forming flimsy cocoons for protection just before pupation. Pupation takes 6 to 10 days.

Cotton leafworm larvae feed on cotton leaves, thus causing a reduction in photosynthetic potential. They can defoliate young plants. Sometimes, the larvae can attack squares and small bolls.

Leafworm populations are regulated by natural enemies, particularly by Trichogramma. Chemical control requires selective insecticides.

Anthonomus grandis (Boll weevil)

The boll weevil was first described in 1843. The boll weevil entered the USA from Mexico in 1892 and it is still a pest of the Americas only.

The boll weevil is a hard shelled insect. On average, the adult weevil is 6-8 mm long and brown to grayish-brown in color with black body edges. Boll weevils have long slender snouts with chewing mouthparts at the tip. The slightly curved snout is approximately half as long as its body. They have two spurs on their front legs, which is a distinguishing characteristic.

The female boll weevil inserts its ovipositor (egg-laying organ) into the cavity formed on the square to lay a single egg and covers it with a sticky substance. Uncovered cavities are feeding sites. Each female has the capability of laying approximately 200 eggs in its life span. Eggs are laid in small bolls, only if squares are not available.

Immature stages of the boll weevil, called grubs, with brownish heads and chewing mouthparts, are found inside squares and bolls. Grubs feed about 10 days before pupating inside squares or small bolls. The pupal stage of the boll weevil is 10-13 mm long, cream colored, with eyes and an obvious snout. In about 4-6 days the pupae turns into an adult boll weevil. The newly emerged adult feeds for about 5 days during which it mates and becomes ready to lay eggs again. The adult boll weevil spends the winter in hibernation, called "diapause," without food, on fence rows, broad leaf plants, litter along creeks, ditch banks and other protected wooded areas near cotton fields.

Under ideal conditions, the life cycle is completed in 16-18 days. Five to eight generations may be produced each year. Overwintered boll weevil adults move to cotton fields just prior to or at the squaring stage. The female hollows out a cavity with its mouth parts and lays eggs inside the bud. In the absence of squares, boll weevils feed on plant terminals. Boll weevil attacks generally begin from field borders, ultimately spreading to the entire field. It damages cotton by feeding and laying eggs in squares and small bolls. Sufficiently sized squares are required for egg-laying so that enough feeding material is available for grubs to develop to the adult stage.

Boll weevil damage is sighted in fields that have flared squares hanging on plants and lying on the ground. Almost all the pollen grains in squares are eaten by the boll weevil, and such squares are bound to shed. First, squares turn yellow and bracts flare/open, and then they fall off the plant.

Chemical control is effective in addition to cultural and biological control measures. The two most successful approaches to control the boll weevil are to minimize the population going into hibernation and to kill the first generation that shifted to the cotton field.

Bucculatrix thurberiella (Cotton leaf perforator)

The adult cotton leaf perforator is a small, slender, usually white moth with black markings on the forewings. The wing span and body length are less than 15 mm.

Eggs are usually laid singly on cotton leaves. Eggs are brownish-white and hatch in about four days. After eggs hatch, early instar larvae make a winding, randomly patterned mine in leaves. The larva is dull amber green with two rows of black spots and distinct white projections on its back. The larvae feed in the mines through the third instar for five days. At the fourth instar they emerge from the mines as free feeders. Damage from these free-feeding larvae affects photosynthetic potential.

The cotton leaf perforator has a characteristic horseshoe-shaped stage on the underside of the leaf that marks the time between the fourth and fifth instar larva. The mature larvae drop to the ground and move to the base of the cotton plants or other plants in the area, and spin a fine web.

The pupa is a cocoon of silk-like material on the stalks of cotton plants. These pupal cocoons are white, approximately 12-15 mm in length and are arranged longitudinally on the plant. The pupal stage lasts for 6-7 days.

Most of the damage occurs before crop maturity and is usually associated with multiple sprays of insecticides.

Diparopsis spp.

Two species of *Diparopsis* are known to have economic significance on cotton: *Diparopsis watersi* (Sudan bollworm) and *Diparopsis castanea* (red bollworm). The Sudan bollworm is found in northern Africa while the red bollworm is common in southern Africa.

Eggs are laid singly or in a group of 2-3 per site. They are blue colored but change to grey before hatching. Eggs are 0.5 to 0.7 mm in diameter and can be seen with the naked eye. Hatched white egg shells can also be seen with the naked eye. Each female may lay up to 200 eggs on the vegetative parts of the plant, and changing to the reproductive parts as the plant grows. Eggs hatch in 6-13 days depending on the conditions.

The larva passes through five instars. The length of the full grown larva is 25-30 mm. The first instar is creamy or grayish-white in color with a black head, legs and spiracles. In the second instar, color changes to more greenish, and from the third instar rose-red arrowhead markings become visible. By the fifth instar, the body color becomes faintly greenish and the head, anal plates and legs become brownish. The developmental period for the five instars is highly temperature-dependent but it is at least 11 days.

The larva falls from the plant to find loose soil for pupation. The pupa is brownish with a greenish tinge in the beginning but develops into dark brown. The pupa is 10-15 mm in length and has a life of 2-3 weeks.

The moth is ready for mating as soon as it emerges. Hybridization between two species is possible due to similar sex pheromones. The moth is 15 mm long and has a wing span of 25-35 mm. The visible areas of the moth have various shades like light brown, brown, deep chocolate, green brown and straw color. The abdomen and hind wings are silvery-cream in color. The forewings are divided by three curved transverse lines into four parts having different colors. In the *D. watersi*, the first transverse line is smooth-curved while in *D. castanea* the same line is sharp-angled and it crosses the median vein.

Larvae prefer to feed on fruiting parts particularly bolls but also damage the shoot if buds and bolls are not available. After feeding, the boll entrance is closed and frass excreted is left inside. Larvae can gain access to bolls if they fall to the ground due to bud shedding.

Spodoptera exigua (Beet armyworm)

The adult beet armyworm has a wing span of 25-30 mm. The forewings are grayish-brown with two yellow spots near the center. The hind wing is translucent white with narrow brown borders.

The larva is approximately 30 mm long with a conspicuous black dot on each side of the second body segment behind the head.

The female lays 500-600 eggs on average in 4-10 days in masses of about 80 eggs beginning early in the spring. These eggs are covered with hairs and scales from her body. The eggs take 2-5 days to hatch.

The larvae feed for about 3 weeks and pass through five instars (growth stages). The small larvae feed in groups for several days but later spread out and become independent feeders.

Beet armyworm pupates in the upper 6-8 mm of the soil in a cell formed by gluing soil particles and trash together with a sticky secretion. The life cycle from egg to adult requires 30 to 40 days, depending on weather. Beet army-worms overwinter in the pupal stage.

The most obvious indication of infestation is the characteristic egg masses coupled with defoliation by larval feeding. Beet armyworms also feed on fruiting forms and occasionally may cause severe damage. Larvae often feed on bracts causing little or no damage.

Spodoptera frugiperda (Black or fall armyworm)

The adult fall armyworm is a dark brownish-gray mottled moth, with oblique markings near the center of the forewing. There is an irregular white or gray patch near the wing tip. Female moths are darker than males. The hindwings are white in females, with a pink luster and a brown border. The wings are about 40 mm in span.

Fall armyworm eggs are similar to those of beet armyworm. They are laid in masses of 50 to 100 and covered with hairs and scales from the female's body. The adult begins laying eggs after 3 or 4 days and lays about 150 eggs a day for 8 to 10 days. Eggs hatch in 3-5 days.

The larva of fall armyworm, when newly hatched, is white with a black head. The color darkens with age. Larvae generally eat all available foliage and then crawl in armies to adjoining fields. Fall armyworm is a general feeder and does not confine itself to cotton. All fruit forms can be damaged, but topping and branch cut off causes the most damage. Larvae have an average age of two weeks.

Young fall armyworm larvae often curl up on a leaf. As larvae mature, they turn greenish-brown with a white line below the top of the back, usually a brownish-black stripe above the midline and a pale stripe with a reddish-brown tinge below. Mature larvae are about 40 mm in length. There is a prominent white inverted "Y" on the front of the head. Fall armyworm larvae can be distinguished from beet armyworm by the presence of black hairs on the body.

Other known species of *Spodoptera* that are important on cotton are *S. littoralis* (leaf worm), *S. litura* (leaf worm) and *S. exempta* (African armyworm).

Sylepta derogata (Cotton leaf roller)

The moth is light creamy or yellowish-white in color. The wings have black borders and numerous brown and black wavy lines. The wing span is 28-40 mm. The head and thorax have black and brown dotted spots while the abdomen has brown rings.

Eggs are laid singly on the underside of leaves. The caterpillar is greenish-white to pink in color and 2-3 cm long. The young larva eats lamina while the damage caused by more advanced stages of the larva is typically in the form of leaves rolled in a funnel shape held inside by silky threads.

The pupa has eight straight spines with hooked tips at its extremity. The life cycle is completed in 23-45 days.

Damage is caused by caterpillars that feed on the lower surface of leaves. As they grow, the leaf edges roll inward up to the mid rib. The attack occurs in patches and may lead to complete defoliation.

Trichoplusia ni (Cabbage looper)

The adult cabbage looper is a grayish brown moth approximately 35 mm long with a wing span of about 40 mm. There is a characteristic silvery spot resembling a "V" or figure "8" with an open end at the top near the center of the forewing. The hindwings are pale brown but darker toward the edge of the wing.

The full developed larva is about 50 mm long and light green with several white lines extending through the length of the body. Larvae have three pairs of slender legs near the head and three thicker, club-shaped prolegs behind the middle toward the rear. The middle half of the body is without legs. Usually, there are two light stripes on the top and one stripe on the side of the body.

The cabbage looper overwinters in a cocoon and usually remains attached to the plant material upon which it has fed. The pupal stage lasts 12 to 14 days before adult moths emerge. The complete generation requires about 35 days.

Adults feed for a short time and begin laying eggs upon emergence in early spring. Cabbage looper eggs are light green and slightly flat. They lack the distinctive ridges of cotton bollworm and tobacco budworm eggs. Eggs are deposited singly on the undersurface of leaves. A single female moth may lay from 200 to 350 eggs in 2 to 3 days. Larvae feed for 2 to 3 weeks before changing into pupae in silk cocoons attached to plants.

Cabbage loopers are mainly foliage feeders. They eat the area between leaf veins, occasionally creating a net-like appearance. Infestations may be greater along field margins. Severe defoliation may occur rarely, which could affect boll maturity, but in general the pest does not reach economic importance.

Bollworms

Cryptophlebia leucotreta (False coding moth)

Cryptophlebia leucotreta, previously known as *Argyroploce leucotreta*, is polyphagous and a serious citrus pest as well. Caterpillars are pink and resemble those of pink bollworm, but differences can be identified under a magnifying glass.

Eggs are about one mm in length, oval in shape and whitish in color. A female can lay up to 400 eggs, mostly in groups, which hatch in 3-6 days.

The full grown larva is 12-18 mm long and orangey-pink in color, with head and first thoracic segment both brown. Larva lives from 10-20 days, depending on temperature. Pupa is an 8-10 mm long, lightly woven cocoon. Life cycle is completed in 35-40 days.

Adults are pale to brown in color. The head is covered with erect brown and black scales. The adult body is about 6-8 mm long with a wing span of 15-20 mm. Males are slightly smaller in span compared to females.

The larva prefers to eat young seeds, so only the late stage of a crop is affected. Larvae enter the medium-age bolls and feeds on seeds resulting in complete loss of the boll. Several caterpillars may be found in the same boll. Like the pink bollworm, the effectiveness of chemical control is low because larvae live inside the carpel.

Cotton **Facts**

Earias spp.

The following species are important on cotton.

Earias insulana	Spiny bollworm (Egyptian bollworm, cotton spotted bollworm, spiny bollworm)
Earias biplaga	Spiny bollworm
Earias vittella	Spotted bollworm
Earias buegeli	Rough bollworm

Distribution is different for different species, and some less important species have also been recorded on cotton. Some species overlap in some countries. The common names given above are not strictly followed.

Most *Earias* spp. eggs are light blue and decorated with approximately 30 longitudinal ridges. A female lays about 90 eggs singly in a lifetime, and the favored site for egg laying is a young shoot. Eggs hatch in about three days. The larva is 13-15 mm in length and has hairs on each segment. The last two thoracic and all abdominal segments have two pairs of tubercles, one dorsal and the other lateral. The tubercles may be long or short and round depending on the species. The base color of the larvae is light brown tinged with grey or green. There are yellow spots on the thoracic segment. *E. insulana* larvae are lighter in color and have a grey and yellow pattern rather than the brown and deep orange in *E. biplaga*. *E. vittella* larvae have less prominent tubercles. The larva lives for about two weeks.

The pupa is yellow to chocolate brown in color and about 13 mm long. The pupal stage lasts about two weeks.

The adult is about 12 mm long and 20-22 mm in wing span with dense and soft scale coating. The abdomen and hind wings are silvery or creamy color. The head, thorax and forewing vary according to species. The primary identification mark among species is the pattern on the forewings, though variations also exist within a species and base color. *E. insulana* has a silvery green to straw yellow base color; *E. biplaga* has a metallic green to gold ground color; and *E. vittella* has a creamy-white or peach base color. There is no diapause in the life history of the *Earias* spp.

The infestation starts with shoot boring in the early stages of the crop. The larva enters the shoot at the terminal/growing point and the injury changes the plant morphology to shorter height and longer sympodial branches. The larva also bores into flower buds and green bolls. The larva does not confine to one fruiting part, rather it moves to other flower buds or bolls leaving the former partially damaged.

Helicoverpa armigera (American bollworm)

The American bollworm, formerly known as *Heliothis armigera*, is also called the cotton bollworm and can be found in all the cotton growing countries in the world.

The caterpillar is initially pale green, sometimes with black dots and a pattern of thin dark lines running along the body, the lines being darker around the second and third segments. In later instars, the dark lines become less conspicuous, and the black spots develop red areas around them. There is a good deal of variation in color in this species.

The caterpillars have a dark triangular area on the back of the first abdominal segment of the third, fourth and fifth instars. The light border patch on the hind wing distinguishes the *Helicoverpa armigera* moth from *Helicoverpa punctigera* moth. When disturbed, caterpillars lift their head and curl it under the body. If even more disturbed, they let go and drop, rolling into a spiral. Larvae do not frequently move from one plant to another. Larval development is rapid under favorable conditions lasting only 19-26 days at 25° C. Fully developed larvae move to the soil where they form an earthen cell 2-10 cm below the surface. A full grown larva is about 4 cms long and pupates in the soil.

The adult moth emerges from pupa in 7-10 days. It has brown forewings with a delicate darker tracery around a single dark mark on each wing. The hind wings are buff with a dark border, which contain a light patch. The undersides are buff with dark sub marginal bands on each wing, and each forewing has a black comma mark and a black dot. Adults are highly migratory and fly long distances.

Mating and egg laying activities are done mostly at night. A single female may oviposit from 500 to 3,000 eggs, averaging close to 1,000. Eggs are usually laid singly and hatch in 3-5 days. Egg color is whitish or creamy white soon after laying but fertile eggs darken in color to grayish brown before hatching.

Young larva feed on leaves and flower buds initially, and then bore into bolls with their head thrust into the boll and the rest of the body outside. The hole is big and round.

Heliothis virescens (Tobacco budworm)

The adult tobacco budworm is similar to the cotton bollworm but with a slightly smaller wing span. The body length is also slightly smaller. The forewings of the tobacco budworm are usually light olive green with three or four light-colored, oblique bands. The adult form is a moth capable of making long flights.

The cotton bollworm passes through four stages: egg, larva, pupa and adult. Eggs are laid on the upper surface of leaves and other tender parts including bracts, blooms, small squares and bolls. Eggs hatch in 2-5 days.

The tobacco budworm and the cotton bollworm look alike in the larval stages. Their size varies from very small when hatched up to 5 cm long as the larva reaches maximum size. The color of the tobacco budworm larva varies from uniform light green to shades of green or brown, usually with stripes running the length of the body. The tobacco budworm larva has a tooth-like projection on the inside surface of the mandibles and fine short hairs on the first, second and eighth abdominal projections which bear a single, prominent spine. A microscope or hand lens is necessary to observe these characteristics. Tobacco budworm pupates in the soil near the host plant in the upper 10-15 cm soil. Adults emerge when weather conditions become favorable and may host on other plants before moving to cotton.

Tobacco budworm adults feed on plant nectaries for 2-5 days before starting egg laying, while larvae feed on fruiting forms for 14-18 days. Life cycle is completed in about 20-25 days but weather conditions have a big effect on the duration of the life cycle.

Newly hatched larvae usually begin feeding on tender cotton leaves and other tender vegetation before attacking fruiting forms. Small larvae feed on small fruiting forms and generally move progressively to larger fruiting forms as they grow. Larvae tunnel into small squares and terminal buds, leaving holes that range in size from very small up to the diameter of a pencil. The entry site is usually at the base of the fruit. Feeding damages or destroys the squares, blooms and bolls. Injured squares flare and drop from plants usually within 5 to 7 days. Large larvae feed on bolls, squares and pollen in open flowers. Bolls damaged by the worm are lost completely, or rot even if not completely damaged.

Heliothis punctigera is limited to Australia. Egg color is the same as in *H. armigera*.

Helicoverpa zea (Cotton bollworm)

The cotton bollworm is also called corn earworm and tomato fruit worm. *Helicoverpa zea* passes through four stages, i.e., egg, larva, pupa and adult. The adult form (moth) is about 3-4 cm in length and capable of making long flights. Forewings vary from light brown or tan to reddish brown and are marked with dark areas near the tip and a dark spot usually near the center. The hindwing of the bollworm moth is white to light tan with an irregular dark band on the outer hind margin. Unlike the tobacco bollworm, the cotton bollworm lacks a tooth-like projection on the inside surface of the mandibles and fine short hairs.

The cotton bollworm and tobacco budworm eggs are similar. The freshly laid eggs are whitish colored and approximately the size of a pin head. Eggs are hemispherical in shape, resembling an inverted cup, with ridges running along the side from the top center to the point of attachment on the plant. They are usually deposited singly on any part of a plant but tend to be placed in the upper third of the canopy. Eggs become tarnish-brown close to hatching. A female can lay from 250 to 1500 eggs during a lifespan of 3-14 days.

The larval color varies from uniform light green to shades of green or brown, usually with stripes running the length of the body. The larval period lasts for 14-18 days. The larvae go into the soil to pupate and adults emerge in 5-7 days.

Adults feed on plant nectars for 2-5 days before starting to lay eggs. The life cycle is completed in about 20-25 days, but weather conditions have a significant effect on the duration of the life cycle.

Larvae tunnel into small squares and terminal buds, leaving holes that range in size from very small up to the diameter of a pencil. The entry site is usually at the base of a fruit. Feeding damages and destroys squares, blooms and bolls. Injured squares flare and drop from plants usually within 5 to 7 days. Large larvae feed on bolls and pollen of open flowers. Bolls will rot even if not completely damaged by the worm. If cotton is not available, the pest may spend some time on alternate hosts like corn and sorghum before moving to cotton.

Pectinophora gossypiella (Pink bollworm)

The pink bollworm moth is grayish-brown with blackish bands on the forewings. The hindwings are silvery-grey. The moth is 12-15 mm in length with slightly shorter wing span. The wing span is 15-20 mm and the wings are folded during resting. Life cycle is completed in 3-4 weeks.

Pink bollworm eggs are greenish-white, turning yellowish and finally orange-red just prior to hatching. The eggs are oval and laid either singly and rarely in small groups of 4-5 on the stem and squares, bolls, leaves and on terminal buds. They are most often found between bracts and the boll wall. Depending on size, each female lays 100-500 eggs in total. Eggs hatch in 4-5 days.

Pink bollworm larvae are white in the beginning but turn to pink with age. The head is yellowish brown, and the full developed fourth instar larvae length ranges from 12-15 mm.

Larvae pupate in the soil during the active cotton season and emerge as moths in about eight days. Larvae generally diapause in leftover bolls at the end of the cotton season.

Larvae feed on the developing flower buds or bolls for about 10-14 days. The newly hatched larva makes its way to the nearest bud or boll, but seeds of 14 to 28-day-old bolls are mostly preferred for food. The affected buds may open as rosette flowers and damaged bolls open partially. The holes where larvae enter into the boll are not visible to the naked eye. Larvae damage may result in complete or partial boll rot. Double seeds, after ginning, are a clear indication of pink bollworm larvae presence.

Pheromones and sterile male moth release techniques have been used in addition to chemicals to control the pest.

Natural Enemies (Beneficials)

The natural enemies of insect pests can be categorized as predators, parasites and pathogens.

Predators – Organisms that kill and consume more than one prey to complete their development and are free living when immature and as adults.

Parasites – Prey on a single host, and may be specific to stages of insects.

Pathogens – Organisms that cause diseases. An organism may be a pest as well as a beneficial organism at the same time (damaging cotton but preying on other cotton insects). Such organisms are sometimes referred to as "benepests."

Predators

Chrysopa spp. (Green lacewings)

Although many species of green lacewings are present in cotton fields, the most abundant are the common green lacewing *Chrysoperla carnea* and the green lacewing *Chrysopa rufilabris*. Adults are yellowish-green with golden eyes and large, delicate netted wings. Common green lacewing adults are not predaceous, but the immature feed on aphids, spider mites, whiteflies and bollworm or tobacco budworm larvae and consume many eggs or small worms.

Geocoris spp. (Big eyed bugs)

Many species have been found to attack cotton pests. The big eyed bug is 4-6 mm in length with a distinctively broad head and large conspicuous eyes. Adults and nymphs feed on mites, whiteflies, thrips, lygus bugs and fleahoppers, while consuming about two bollworm or tobacco budworm eggs per day under field conditions. The big-eyed bug may also feed on plant sap without any damage to the plant.

Hippodamia spp. (Lady beetles)

The most important species is *Hippodamia convergens*. The adult is generally orange in color with 12 black spots on the wing covers. The orange and black grub is predaceous. Lady beetle populations are related to aphid numbers, so lady beetles are most abundant when aphid populations are high. The lady beetle may eat bollworm eggs and also attack small bollworm larvae.

Nabis spp. (Damsel bugs)

Several species of damsel bugs are found on cotton and one of the most abundant species is the pale damsel bug (*Nabis capsiformis*. Adult damsel bugs are slender with relatively long legs. The color ranges from tan to reddish-brown. Damsel bugs feed on bollworm-budworm eggs and larvae, as well as on aphids, fleahoppers, lygus bugs, leafhoppers and spider mites. They also prey on other predators such as minute pirate bugs and big-eyed bugs.

Orius spp. (Minute pirate bugs)

Two species are found in cotton fields, *Orius insidiosus* and *Orius tristicolor*. Both are black and white measuring about 3-4 mm in length. They feed on aphids, thrips, whiteflies, mites and bollworm-budworm eggs and small larvae.

Pseudatomoscelis seriatus (Cotton fleahoppers)

The cotton fleahopper is a cotton pest because it causes small squares to shed. However, the fleahopper also serves as a predator on bollworm or tobacco budworm eggs. Fleahoppers are generally considered pests on preflowering cotton, but they are considered beneficial after flowering. Both adults and nymphs are predaceous but also serve as food for many other predators.

Solenopsis spp. (Ants)

Many species of predaceous ants may be found on unsprayed cotton, particularly if sugary insect secretions are available. Although individual feeding efficacy is very low, their high population makes them effective predators.

Zelus and Sinea spp. (Assassin bugs)

The most abundant assassin bugs in cotton are the leafhopper assassin bug *Zelus renardii* and the spined assassin bug *Sinea diadema*. Leafhopper assassin bug adults have various colors including red, brown and yellowish green. They are predators of cotton fleahoppers and bollworm-budworm eggs and larvae, and are one of the few predators that can readily capture and consume a boll weevil adult as well as large *Heliothis* larvae.

Catolaccus grandis

Catolaccus grandis is a medium size wasp with large eyes and short antennae. It is a native of Mexico and parasitizes boll weevil grubs in the third instar. The wasp searches squares for boll weevil grubs, paralyzes the grub by stinging and places an egg on it. The egg hatches and the parasite larvae feed on the paralyzed grub for many days. The full grown larva pupates inside the square or boll and the adult wasp emerges in about five days. The wasp life cycle is completed in 12-15 days under field conditions. Efforts have been made to rear the wasp under lab conditions but the high cost of rearing remains a concern for commercializing its use.

Spiders

Spiders are important in regulating insect pests but are not numerous enough to control excessive numbers that occur during major outbreaks. Along with other natural predators and parasites, spiders can keep pests below unacceptable levels.

Acanthepeira stellata (Star-bellied orb weaver)

The star-bellied orb weaver has a star shaped abdomen formed by a series of cone-like bumps. The spider is brown in color with a white spot on the anterior portion of the abdomen. Adult fleahoppers, bollworm-budworm moths, big eyed bugs, pirate bugs, damsel bugs and honeybees are frequent prey of this spider.

Aysha gracilis (Grey dotted spider)

The adult spider is about 6-8 mm long and yellow with a darker anterior end and brown jaws. It has a pair of longitudinal grey bands between the eyes and the abdomen, and reddish-brown to black spots form two indistinct longitudinal bands on the abdomen. The spider preys on bollworm-budworm eggs and small worms.

Chiracanthium inclusum (Winter spider)

The winter spider is found in silken nests under leaves or bracts of squares. The adult spider is 8-10 mm long and cream to pale green in color. Eggs and first instar worm larvae are the primary targets but may also prey on bugs, grasshoppers and beetles.

Misumenops celer (Celer crab spider)

The celer crab spider is generally found in the plant terminal parts. It has eight legs, the front four legs are longer than the four rear ones. There are two longitudinal stripes behind the eyes and the rear half of the abdomen often has a "V" shaped mark pointing toward the rear. An adult is about 6-8 mm long. Celer crab spiders feed on cotton fleahoppers, lygus bugs and a wide range of other insects, including bollworm-budworm eggs and worms.

Oxyopes salticus (Striped lynx)

The striped lynx is a brown-colored 6-10 mm long spider with four longitudinal grey stripes behind the eyes on the front half. Two black lines extend down the face and the jaws. The striped lynx preys on cotton fleahoppers, lygus bugs and small bollworm-budworm larvae and rarely on bollworm eggs.

Phidippus audax (Black and white jumping spider)

The adult spider is black in color with three white spots on the abdomen. The adult is 12-15 mm long. The spider may feed on bollworm-budworm eggs and all larval stages of bollworm-budworms. Boll weevil adults and bollworm-budworm moths can also be attacked. Immature spiders capture and consume fleahoppers, lygus bugs and pink bollworm adults.

Insect Parasites

Insect parasites feed internally or externally on their hosts. The host becomes paralyzed before it dies from the parasite's feeding. Insect parasites can attack any life stage of their host, but most parasitize eggs or larvae.

Trichogramma spp.

The tiny wasps called *Trichogramma* are typical insects that parasitize bollworm and budworm eggs. The adult female wasp deposits eggs inside the bollworm egg by using her ovipositor, and place one or more eggs inside the host. The wasp's egg hatches into a larva which feeds of the bollworm's egg. The full grown *Trichogramma* larva transforms into the adult wasp, which escapes from the dead bollworm egg. The female wasp mates and flies in search of other bollworm eggs to parasitize, thus repeating the life cycle. Trichogramma also parasitizes other moth eggs.

Many species of *Trichogramma* are known parasites of many bollworms, budworms, loopers and other caterpillar pests. These include *T. pretiosum, T. brasiliensis, T. pintoi* and *T. platneri*. The proper procedure and frequency of release may vary considerably, depending on the target caterpillar species, their density, crop habitat, and cultural practices in use. Trichogramma pintoi has been reared in Uzbekistan for a long time and has been commercially used on cotton.

Eretmocerus spp. and Encarsis spp.

        Eretmocerus spp. and *Encarsis spp.* are the most important natural enemies of whiteflies and both parasitize nymphs. The whitefly nymphs parasitized by *Encarsis spp.* turn black while those parasitized by *Eretmocerus spp.* do not. A hole in the head of the whitefly nymph shows that they are parasitized. Adult wasps also act as predators by feeding on whitefly body fluid. A single wasp can parasitize many nymphs.

Pathogens

        Many forms of pathogens have been used to control insect pests and mites. The most used agents are *Bacillus thuringiensis,* nuclear polyhedrosis viruses, etomophthora aphidis, and fungi like *Beauveria bassiana*, *Erynia* spp. and *Nomouraea rileyi.*

Insecticide Resistance

        The cotton plant is attacked by a number of insects and many of them require more than one insecticide spray for effective control. Multiple applications of the same insecticide year after year enable target insect populations to develop resistance.

What Is Insecticide Resistance?

        The World Health Organization has defined resistance as "the development of an ability in a strain of insects to tolerate doses of toxicants that would prove lethal to the majority of individuals in a normal population of the same species."

        Cases of development of resistance were first reported in the USA in the 1950s. Some insects develop resistance sooner than others and, similarly, some insecticides/products are more prone to develop insect resistance. Pest control in the presence or absence of pesticide resistance should include tactics that minimize (further) selection of the target population by the pesticide. Several operational, biological and genetic factors impact upon the potential for insecticide resistance development. Some of these factors are:

        Insect characteristics – such as mobility, host range and diversity, number of generations, time to complete development and reproductive capacity.

        Insecticide characteristics – such as mode of action, similarity to existing products longevity of resistance and susceptibility to metabolism or other degradation.

        Insecticide use – Repeated use of insecticides on various stages of a particular insect produces repeated selection.

Measuring Resistance

Resistance may vary in quality as well as quantity, so it is important to measure the level of resistance for which various methods are employed.

Bioassay techniques – A number of bioassay methods are used to measure resistance, which include topical application, immersion technique and residual bioassays.

The residual bioassay technique could be employed in the form of leaf dip, leaf spray, vials (glass vials coated with residual film of an insecticide), sticky card technique and dietary. Some of these techniques are specific to insecticide exposure to the insect.

Biochemical tests – Resistance can also be measured by biochemical tests that measure or identify specific enzymes that confer resistance. The additional advantage of the biochemical technique is that tests can be carried out on small samples.

Other techniques like diagnostic dose or bioassays have also been employed in resistance management, particularly in the case of changes in the resistant gene frequencies.

LD50 (lethal dose 50) and LD90 (lethal dose 90), are doses of insecticides at which 50% and 90% of the population will be killed.

Why Insects Develop Resistance (Resistance Mechanisms)

Insects develop resistance due to the following:

Reduced penetration through cuticle.

Metabolic mechanisms – Insects can develop the ability to metabolize and excrete toxic chemicals, as has been the case with most organophosphates, carbamates and pyrethroids.

Insensitivity of the target site (nervous system).

Genetic resistance – Insects develop insecticide-resistant genes that are transmitted to the target population.

Synergists can be involved in two ways, inhibition of detoxifying enzymes and counteracting metabolic resistance to increase toxicity.

Cross resistance – The ability of insects to tolerate higher doses of a particular chemical or chemicals due to development of resistance to other chemicals.

Multiple resistance – Insects can also develop resistance to more than one mechanism at the same time.

Insecticide Resistance Management Guidelines

Do not rely heavily on insecticide use and always employ all the efficient cultural/biological control practices available in the pest control system.

Do not rely on a single insecticide class for a long time.

Always time insecticide applications during the most susceptible life stage of the insect based on the local pest threshold.

Use insecticides only at recommended doses and at recommended spray intervals

When choices are available, use different classes of insecticides in alternation.

If an insecticide fails to control the target insect at the recommended dose, **do not** try a higher dose of the same insecticide.

If insecticides are mixed together and sprayed to control target insects, each insecticide should be mixed in the proper recommended dose.

INSECT CONTROL METHODS

Control measures can be placed in seven different categories. They are:

Biological control – Biological control includes utilization of predators, parasites, pathogens and microbial agents. Biological methods include encouragement of existing natural enemies and introduction of artificially reared biological agents.

Host plant resistance – Host plant resistance includes in-built genetic resistance through special plant characters like frego bract, red plant color, nectari-less leaves (no gland-like organ on back of the leaf), leaf hairiness, glabrousness, high gossypol, okra leaf, early fruiting, etc.

Cultural control – Cultural control methods include crop rotation, stalk destruction and diapausing sites on plants and soil, least variation in the fruiting habit of cotton varieties, growth regulators and agronomic practices like proper weeding, optimum fertilization, irrigation, least variation in planting time, proper plant spacing, trap cropping, etc.

Legislative control – Legislative control includes quarantine laws, control of plant movement, enforced rotation, destruction of crop residues by law, official zoning of varieties, certified quality seed production and distribution, etc.

Physical control – Physical control includes hand-picking of eggs/larvae and brooms/branches to beat pests.

Special control – Special control methods may be specific for specific pests like male sterility techniques, repellents/attractants, pheromones etc.

Chemical control – Cotton insects and mites can be controlled chemically with natural or synthetic insecticides and ovicides. The major classes of synthetic insecticides used to protect cotton are carbamates, chlorinated hydrocarbons, organophosphates, pyrethroids and new chemical compounds. The development and use of synthetic insecticides has not only increased yields but also made it possible to produce cotton in areas where it would not otherwise be possible to grow and harvest with economical yields.

Although insecticides can have severe adverse effects on agro-ecosystems, they are quick and more effective than other methods. The potential for adverse effects requires judicious use of toxic chemicals.

Chemicals have been used for pest control for centuries. Insecticides can be systemic (soluble in water—reaching insects through cell sap), contact (stomach poisons) and respiratory in action.

There are many different types of chemicals used for pest control. The historical development of insecticides has proceeded as follows:

1. Inorganic insecticides
2. Botanical insecticides
3. Modern synthetic chemicals
   A: Chlorinated hydrocarbons
   B: Organophosphorus
   C: Carbamates
   D: Pyrethroids
   E: Insect growth regulators
   F: Formamidine insecticides

Inorganic/Naturally Occurring Insecticides

Materials like salts, ashes, soot, dust and sulfur have been used for pest control for hundreds of years. Salts of mercury and lead were also used.

1600s: Arsenic + honey was used as stomach poison for ants.

1800s: Paris Green (arsenical) was used to control Colorado potato beetle and other foliage feeding insects.

1920-40s: Calcium arsenate was developed which was a stomach poison inhibiting respiratory enzymes.

Use of inorganic/naturally occurring chemicals led to numerous problems: aerial dusting to control the boll weevil resulted in outbreaks of secondary pests due to destruction of natural enemies. Arsenic in the soil led to low yields in arsenic-sensitive crops like soybeans.

Botanical Insecticides

Nicotine (water extract of tobacco) has been used since 1763. Nicotine worked as a contact nerve poison (agonist of nicotine acetylcholine receptors), but the major drawback has been high human fatalities.

Pyrethrum (extracted from chrysanthemum flowers) was a contact nerve poison that disrupts sodium channel function. It had a fast knockdown effect but was highly photosensitive and bio-degradable. Pyrethrum was used extensively during the 1800s.

Others include derris, rotenone, sabadilla, ryania, hellebore, etc.

Modern Synthetic Chemicals/Insecticides

Chemical composition and action on the plant differentiates insecticides into various groups. There are chemicals/insecticides developed for specific objectives, but major insecticides are divided into the following groups:

Chlorinated hydrocarbons

Paul Hermann Müller described the insecticidal properties of Dichloro-Diphenyl-Trichloroethane (DDT) for the first time in 1939. DDT was a nerve poison with contact and stomach actions and a broad spectrum insecticide with a long residual effect, and hence offered economical pest control. Paul Hermann Müller was awarded a Nobel Prize for his discovery in 1948.

1962: Over 1200 formulations of DDT against 240 pests had been developed.

1963: DDT production increased to 81 million kilograms

1972: DDT was banned in most countries.

Organophosphates

G. Schrader discovered schradan in 1941, a nerve poison inhibiting acetylcholinesterase, an enzyme in the nervous system. Parathion was developed in 1944. Other famous organophosphates are malathion, dimethoate, diazinon, and profenofos. The mode of action is contact, stomach, systemic and fumigants.

Organophosphates – Comparison with hydrocarbons

	Organophosphates	Chlorinated HCs
Persistence:	Low	High
Biodegradability	High	Low
Mammalian toxicity	Higher	Relatively lower

Organophosphates and hydrocarbons are similar in effects on pest status, resurgence of target species, destruction of beneficial species and development of resistance.

Carbamates

Carbamates are relatively non-persistent and, unlike organochlorines (chlorinated hydrocarbons), they do not bioaccumulate. They are synthetic analogues of plant alkaloid physostigmine and the route of absorption could be ingestion, inhalation and dermal.

Carbaryl, a broad spectrum and low mammalian toxicity product, was developed in 1956. Other carbamates include aldicarb, methomyl, thiocarb, propoxur and carbofuran.

Synthetic pyrethroids were developed in 1972. Permethrin (synthetic and photostable) was effective at rates of 10-100 times less than organophosphates and chlorinated hydrocarbons. Permethrin was an axonic nerve poison, similar to DDT in its mode of action. Therefore, cross resistance was a potential problem. Pyrethroids have been extensively used on cotton. Other forms of pyrethroids are fenvalerate, cypermethrin, bifenthrin etc.

Insect growth regulators

The structure of the first juvenile hormone was elucidated in 1967, but insect growth regulators have a slow mode of action and so they are not popular. The mode of action can be different, as with methoprene that kills by inhibiting molting benzoylphenylureas – interferes with chitin synthesis.

Formamidine insecticide

So far, only two commercially successful formamidines: amitraz and chlordimeform have been approved. The mode of action is ovicidal via contact and vapor, synergism of many classes of insecticides, and modification of insect behavior. Sublethal effects include disruption of feeding and reproduction, functioning as octopamine agonists.

Pesticide Classification

The World Health Organization (WHO) classifies pesticides on the basis of their toxicity. This classification is as follows:

WHO Classification: LD50 for rats (mg/kg body weight)

WHO Class	Oral		Dermal	
	Solids	Liquids	Solids	Liquids
Ia Extremely hazardous	5 or <	20 or <	10 <	40 <
Ib Highly hazardous	5 – 50	20 – 200	10 – 100	40 - 400
II Moderately hazardous	50 – 500	200 – 2000	100 - 1000	400 - 4000
III Slightly Hazardous	Over 500	Over 2000	Over 1000	Over 4000

LD50: Statistical estimate of number of milligrams of toxicant per kilogram of body weight to kill 50% of a large population of test animals

There is a fifth group of chemicals which is considered safe.

COMMONLY USED TERMS IN INSECT PEST CONTROL

Antagonists. Organisms that release toxins or otherwise change conditions so that activity or growth of other organisms, particularly pests, is reduced.

Augmentation. Augmentation is the periodical release of biological control agents for natural control of pests.

Beneficials. In cotton production systems, beneficial usually applies to natural enemies of pests and to pollinators such as bees.

Biological control. The use of parasites, predators or pathogens in maintaining another organism's population density at a lower level than would occur in their absence. Biological control may occur naturally in the field or in lab-rearing of specific biological control agents against specific target pests and then their release in the field.

Broad-spectrum pesticide. A pesticide that is able to kill many different species of pests at the same time.

Economic threshold. Economic threshold is either a level of pest population or damage at which the cost of pesticide use at least equals the savings in the crop value gained from control action.

Instar. The larval or nymphal stage of an immature insect between successive molts.

Integrated pest management (IPM). A pest management strategy that focuses on long-term prevention or suppression of pest problems by employing a combination of all possible means like biological, agronomic, host plant resistance, chemical, etc., rather than relying only on one control measure.

Lay-by application. Application of any input like a fertilizer, herbicide or insecticide on a specific area of the field/plants after the crop is well established.

Multi-adversity resistance. A combination of several resistance mechanisms in a single genotype.

Nymph. The immature stage of insects that hatch from eggs and gradually acquire adult form through a series of molts without passing through a pupal stage.

Oviposition. The laying or depositing of eggs.

Parasite. An organism that lives in or on another usually larger organism (host) without killing the host directly. Also an insect that spends its immature stages in the body of a host that dies just before the parasite emerges (this type is also called a parasitoid). Parasites of insects, termed parasitoids, are a special group of parasites since they kill their host.

Parthenogenesis. Development of an egg without fertilization.

Pesticide resistance. The ability of an organism to survive a pesticide application at doses that once killed most of a population. The ability is usually acquired after excessive and continuous use of pesticides.

Predator. With respect to cotton insects, a predator is an insect that preys on other insect pests as a source of food.

Pheromone. A substance secreted by an organism to affect the behavior of the opposite sex of the same species. Sex pheromones that attract the opposite sex for mating are used in monitoring insects. Pheromones have also been used as a control measure against certain insects especially Pectinophora gossypiella in cotton.

Phytotoxicity. The ability of a material such as a pesticide or fertilizer to cause injury to plants.

Pupa. An insect between the immature and adult stages, which generally does not feed but is capable of undergoing complete metamorphosis.

Rosette flower. A flower that has petals tied together with silk by the pink bollworm larva.

Secondary outbreak. The increase of a non-target pest to harmful levels due to an imbalance in natural biological control, usually emerging from pesticide applications.

Solarization. The practice of heating soil to lethal levels to kill pests through application of clear plastic to the soil surface for weeks during sunny, hot weather.

Systemic insecticides. Insecticides capable of moving from roots throughout the plant, usually in the vascular system. In cotton, systemic insecticides are used to control sucking insects.

Fiber Quality

Cotton is a cellulosic fiber about 96% pure. Cellulose is a naturally occurring carbohydrate polymer, which forms the basic raw material for the production of rayon and acetate. Fibers such as nylon, polyester and acrylics made from petrochemical raw material are called non-cellulosic fibers.

Cotton was one of the first agricultural commodities to be sold on the basis of quality. Distinctions were first made between species and then between growths within species. The mix of a perennial crop and annual growth habits contribute to the variability in cotton lint quality.

The amount of sunlight, day and night temperatures during growth, variety and agronomic inputs are responsible for year-to-year variations in quality. The use of bales of uniform fiber properties or the blending of bales will contribute to consistent processing and uniformity of yarn.

Fiber properties have been studied since the early 1900s, but electronic and physical sciences have been employed in measuring quality parameters only since the 1950s. High volume instruments (HVI) are machines for measuring quality characteristics in cotton. The purpose of developing high volume instrument systems was to automate functions and reduce the time required to measure fiber properties.

Fibers competing with cotton include natural fibers such as wool, silk, alpaca, jute, hemp, flax, ramie, abaca (Manila hemp), bamboo, kenaf and sisal. Manmade fibers include acetate, viscose, polynosic, acrylate, acrylic, modacrylic, nylon, olefin, polyester, polyolefin, polypropylene and carbon.

Cavitoma is a term used to describe cotton damaged by microbes.

FIBER FORMATION

Each cotton seed is capable of producing up to 20,000 single hairs/fibers. Formation of fiber hairs on the seedcoat is not unusual, but the characteristics of such hair are unique and largely dependent on species. Each cotton fiber is a single cell emerging from the seed coat. Cotton fiber grows from the epidermis of the seedcoat cell. Fibers begin to elongate from the day of anthesis (flowering) and this growth is completed in 15-25 days.

Fibers consist of a primary wall and a secondary wall, both mainly comprising cellulose fibrils. The primary wall is covered with a cuticle consisting of mainly wax, pectinous substances and reducing sugars. Entomological sugars, if any, are deposited on the cuticle (upper surface of fiber) and may cover the natural wax.

In the primary wall, fibrils are placed in a wide angle with the fiber axis, and thus the primary wall has no effect on fiber strength. In the secondary wall, the cellulose fibrils are laid side by side following a helical course, making a small angle known as the spiral angle. Thus, the secondary wall thickening has a significant effect on fiber strength. The spiral angle in varieties differs greatly. As a general principle, a smaller spiral angle forms steeper spirals and thus stronger fibers. The spirals occasionally reverse direction resulting in structural reversals, which are potentially weak points in the fiber.

The secondary wall forms more than 90% of the fiber weight. Fiber strength is directly related to yarn strength. Strong fibers make a strong yarn. However, this relationship is affected by fineness, length and length uniformity.

The diameter of fiber is reached soon after it originates, and thus intrinsic fineness is established at an early stage and remains constant afterward.

Elongation of the single cell forming the fiber continues for 15-25 days, depending on species as well as variety. This is the time when fiber length and length distribution are established. Once the fiber length is established and it ceases to grow, the second stage of fiber formation takes place during which cellulose is deposited in successive layers on the inner surface of the primary wall. Cellulose deposition is completed in about 25-35 days.

When the boll opens, the fibers dry, losing their cylindrical cross section into a flat ribbon shape with many convolutions along their length.

Effects of acids

Cold concentrated acids or hot dilute acids hydrolyze cellulose and disintegrate cotton fibers. Weak cold acids are unable to initiate hydrolysis.

Effects of alkalis

Cotton shows excellent swelling (mercerization) in caustic, but no damage.

Effect of heat

Cotton has high resistant to thermal deformation and degradation but may turn yellow after five hours at 248°F.

Dead cotton

An extremely immature cotton, having a thin fiber wall, is called dead cotton.

FIBER QUALITY PARAMETERS

Elongation

Cotton fiber is flexible and can be stretched. The increase in the length or deformation as a result of stretching is called elongation. Elongation is measured as percentage increase over the unstretched original length.

Length Measurement

Length is measured and expressed in many ways. Some units of length are as follows:

Staple Length. An estimate of length made by personal judgment of the appearance of a hand-prepared staple/tuft. Staple length is the normal length measured by a classer on a typical portion of fibers. A sample of fibers is pulled from cotton and by a process of lapping, pulling and discarding, the fibers are made parallel for measuring the length.

Effective Length. The upper quartile of a numerical length distribution from which some short fibers have been eliminated by an arbitrary construction.

Mean Length. Also called average length, is the arithmetic mean of lengths of all the fibers in the test specimen based on weight-length or number-length data.

Modal Length. The length in a fiber length distribution frequency diagram which has the highest frequency of occurrence.

Upper-quartile Length. The fiber length, which is exceeded by only 25% of the fibers by weight in the test specimen when tested by the array method.

Upper-half Mean Length. The average length of the longer half (50%) of fibers in a sample. The HVI machine gives UHML as the average length by number of the longer half (50%) of the fibers distributed by weight.

Span Length. The distance spanned by a specific percentage of fibers in a test beard when tested by the fibrograph, taking the amount reading at the starting point of scanning as 100%.

2.5% Span Length. A length that is crossed by only 2.5% of the beard fibers scanned by the fibrograph. The 2.5% fiber length is closely associated with the classer's length.

50% Span Length. The length spanned by 50% of the fibers in a beard when scanned by the fibrograph. The 50% span length is important because it gives an indication of the uniformity of length of fibers in the specimen.

The *Suter-Webb Sorter* method involves the use of a series of combs with which fibers are separated in length increments of 1-6 mm. Fibers in each length group are then weighed to determine the weight-length distribution from which length indices can be calculated.

Length Uniformity Index. The ratio between mean length and upper-half-mean length expressed as a percentage of the latter.

Length Uniformity Ratio. The ratio between two span lengths as a percentage of longer length. The 50% and 2.5% span lengths are usually used.

Micronaire. Micronaire was introduced in 1946 as a measure of cotton fineness. Micronaire measures a combination of fiber fineness and maturity. Micronaire is indirectly determined according to the airflow principle. A mass of coarse fibers permits more airflow and thus expresses higher micronaire value. Less airflow is an indication of finer fibers. Repeatability for the micronaire data is high and for a given variety, it is an index of maturity. For unknown varieties, it could be an incorrect gauge for fineness and maturity. Low micronaire values could be an indication of low intrinsic fineness (diameter) or low maturity. The airflow measurement of micronaire by Fibronaire is done on a 3.25 gram sample.

The airflow system to measure micronaire was developed in the late 1940s; however, the airflow micronaire readings were standardized in 1957. Low micronaire cottons produce more neps. They also produce dyeing irregularities and promote spinning waste.

G. barbadense varieties have the lowest micronaire value among cultivated species.

Cultivated diploids have higher micronaire values compared to cultivated tetraploid species.

Among upland cottons, lint with micronaire below 3.5 is usually considered immature and weak.

Maturity. The term maturity refers to the degree of development or thickening of the fiber cell wall relative to the perimeter or effective diameter of the fiber. Lower temperatures than normal during fiber development result in lower maturity.

Fineness. Fiber fineness can be expressed in terms of linear density or weight per unit length. Fineness can be measured by sorting fibers on the Suter-Webb apparatus and then weighing a specific number of fibers.

Intrinsic fineness (perimeter) is a varietal character little affected by environmental factors.

Fiber Color. It is important to measure the color of cotton for at least two reasons: color is an indication of fiber quality, and variations in the color of raw cotton cause variations in the color of dyed cotton products.

Fiber color is expressed practically with two indices, +b and Rd. These values are obtained by means of the cotton colorimeter.

Hunter developed an indicator, known as +b, of the degree of yellowness while Rd refers to reflectance, which is the degree of whiteness of the sample.

Tensile Strength. The maximum load per unit area of the original cross section obtained prior to rupture. It is the actual number of pounds of resistance that a bundle of fibers or fabric will give to a breaking machine before the material is broken.

Fiber strength is related to the average length of the cellulose molecules deposited inside the cotton fiber. The longer the cellulose chains, the stronger the fiber.

To improve fiber strength through breeding, it is necessary to induce steeper spirals and lower reversal frequency. But the complexity of these characteristics is not understood, and hence genetic manipulation of the plant for fiber strength has not yet been accomplished.

Neps. Small knots of entangled fibers. In cotton, neps are usually formed by dead or immature fibers that are not properly straightened during carding and combing. However, in spite of good carding and combing, seed coat fragments and stickiness can cause neps. The chances of increasing the number of neps in cotton are greater with each processing operation.

Seed Coat Fragments (SCFs). SCFs are contaminants in cotton lint. The propensity to produce SCFs is a varietal character. Harsh ginning can increase their size and number. Growing conditions also have an influence. The size of fragments detached from the seed is also variety-dependent, but the same variety grown at various locations following the same set of production practices may give different quantities of seed coat fragments.

Motes. Three possible components of motes are unfertilized ovules, aborted and undeveloped seeds and small seeds. Any factor, genetic, agronomic or climatic, that adversely affects fertilization and seed development has an effect on mote formation.

Short Fiber Content. Fibers shorter than 12.7 mm (half an inch) are called short fibers. However, while this is the most used definition, it is not the best definition.

Short fiber content is expressed as the percentage of the weight or number of fibers in a sample.

There are two sources of origin of short fibers: natural occurrence and origination during processing.

Excluding fuzz, which is not removed from the seed coat during ginning, the quantity of naturally occurring short fibers is smaller compared to the quantity created during processing (plus picking in the case of machine picking).

Variety, growing conditions and their interaction determine the natural occurrence of short fibers on the seed.

Fibers break during processing, and most short fibers originate during ginning. Immature, thin-walled fibers are more prone to breakage during processing.

Floating Fiber Index. The floating fiber index is also an expression of short fiber content usually defined as the percentage of fibers which, in a drafting zone, are not clamped by both rollers of the drafting system. All fibers, short or long, are floating for some time, but for longer fibers this time is very short. Short fibers, on the other hand, float for a longer time, sufficient enough to be detrimental to the drawing process.

Grading

The first quality distinctions, which ultimately emerged into the modern day classification and grading, were based on the area of cultivation and type of cotton (either *G. hirsutum* or *G. barbadense*). It was not until the 19th century when specific quality terms started to be used when referring to quality. Upland cotton standards were promulgated in the USA in 1914. The official U.S. cotton standards are also called Universal Cotton Standards. The Universal Cotton Standards Agreement was signed in 1924. Many companies in many countries recognize the universal cotton standards but classification systems vary among countries. In 1992, the U.S. Department of Agriculture started classing all cottons using HVI.

Grade is the overall appearance of a sample primarily based on a classer's assessment of color, visible trash and preparation (neppiness, rough/smooth ginning). Classers use standards boxes as an aid to class cotton. *G. hirsutum* and *G. barbadense* have different grade standards.

Fiber samples are taken during ginning, at the time of packing and wrapping of bales. After the samples are drawn, care must be taken in handling them so that trash is not lost and the original nature of the sample is preserved for accurate classification.

Color, the principal parameter used to grade cotton, changes if cotton is exposed to weather in the field for a long period of time.

Trash is undesirable. However carefully the cotton may be picked, some trash, particularly leaf trash, is always picked along with cotton under commercial picking conditions. More trash means more cleaning cycles to ensure an acceptable level of purity, but more cleaning causes more damaged fibers. Most trash is removed during pre-cleaning and post-cleaning operations at the gin, but it is impossible to eliminate all trash. It is easy to remove larger particles of trash. Field loss and trash could be equal to zero in hand picked cotton, if adequate care is taken. Cleaning machines eliminate larger trash particles but the quantity of peppery trash increases.

MACHINE PICKING

Cotton is harvested by hand picking or by machines. Regardless of picking method, boll opening in the shortest possible time is desirable for the best picking results. There are only two major kinds of seedcotton harvesters: pickers (spindle harvesters) and strippers.

Pickers were introduced in the USA in 1942. With pickers, harvesting is accomplished by revolving spindles or straight spindles. The spindles may have one, two, three or even four rows of machined teeth depending upon the shape of the spindle. Multiple columns of rotating spindles are arranged around a rotating drum, which projects the spindles toward the plant. The spindles are withdrawn from the plant with seedcotton wrapped around them. The rotatory or stripped doffer brushes away the seedcotton from the spindles. The seedcotton thus removed is collected in a basket in the back of the picker.

The stripper mechanism strips the entire plant of open, semi open or even unopened bolls. The stripping operation may be based on the finger principle, or the plant may be made to pass through rotating rollers.

The major differences in the two methods are the percentage of trash and time needed for harvesting. Trash is much higher in stripped cotton compared to spindle picked cotton. On average, picker-harvested cotton has about 6% trash and about 11% field loss. The picker head can be adjusted to minimize the collection of trash and soil. Stripper-harvested cotton may have up to 25% trash, but the field loss is as low as 2-5%. Adjustments can be made in the stripper to reduce foreign matter. Spindle picked cotton has fewer neps, lower short fiber content, longer staple length, better length uniformity and higher strength compared to stripped cotton.

Stripping is done only once a season, while spindle picking can be repeated in the same field if there is a need to do so. A stripper harvester can cover more area per hour than a picker harvester. The time it takes to harvest one hectare with a picker depends on yield, field conditions, machine conditions, field length and shape, weather and the machine operator. A four-row picker can harvest one hectare in 1-1.5 hours. The first pick takes longer than the second pick.

Machine harvesting requires specific cultural practices in order to be successful. These include planting pattern and plant size, weed control, and defoliation.

Cotton has to be planted in rows for machine harvesting. Proper weeding is important for machine picking of seedcotton. Taller varieties and fruiting too close to the ground are not desirable for machine picking. Sympodial branching is more desirable for machine picking compared to plant types with many monopodia. Cotton from weedy fields will have lower grades compared to cotton from fields free of weeds. In stripper harvesting, tall weeds choke the machine by entangling fingers and rolls.

Defoliation is a prerequisite for machine harvesting. Defoliants should be applied when 60-75% of the bolls have already opened. Application of defoliants prior to this stage not only reduces yield but also forces premature boll opening resulting in immature fibers. The hastening of boll opening, leaf shedding and shedding of small immature bolls (that are not considered to be productive bolls) can also be achieved by applying desiccants.

Module System. This is a system to store seedcotton at high density. All operations are mechanical, and the average density is about 200 kg/m^3 of seedcotton. A module is a big compressed block of seedcotton. Large scale farming operations and machine picking require module storage. Modules are usually built on the ground and can store up to 15 tons of seed cotton.

Cotton Caddy System. This is a system to store seedcotton where seedcotton is directly received from the harvester but it is pressed manually. Thus, density is low.

Mechanical Ricker. In this system of storage, seedcotton is collected from a harvester and a stack is formed along the turnrow in the field. A mechanical ricker has a density of almost 50% of a module.

GINNING

Cotton, when harvested from the field, has lint and seed. The unginned cotton is called seedcotton. Ginning is the process of separating lint from seed by mechanical means. Cotton must be ginned before it is processed in a spinning mill. Originally, ginning was performed by hand by pulling fibers from seed. In one day, one person can remove about one kilogram of lint by hand. Over the years a number of types of ginning machines have developed: the baelna, the roller gin and the saw gin. Prior to the invention of the saw gin, cotton production worldwide was hampered due to the lack of an efficient ginning system.

The *baelna* is a primitive type of roller ginning machine used in the Indo-Pak subcontinent in olden days wherein lint was pulled from seed with the help of two counter rotating rollers. The currently used baelna has one roller made of wood with a diameter of about two inches and the other roller is made of smooth metal with a diameter of less than an inch.

The *roller gin* with reciprocating knife was developed by Fones McCarthy in 1840. In the roller ginning system, fibers are caught between the roller and the fixed knife and pulled from the seed while a moving knife knocks off the seed. The distance between the rollers and the knife does not allow seeds to pass through with the lint, but trash and motes do make their way through.

The main objective in roller gin research has been to improve the efficiency of ginning. By 1880, most hand labor operations had been replaced by mechanical screw processes, gin feeders and pneumatic cotton handling systems. The rotary knife type roller gin was developed during the 1960s.

The *Saw Gin* was developed by Ely Whitney, who saw people removing lint from seed by hand and came up with the idea of removing lint by machine. He developed a saw gin in 1793. The gin had spikes that were later replaced with circular saws by Hodgen and Holmes in 1796. Whitney received a patent on the saw ginning method, and with his colleague, installed gins in the USA. They charged a fee as high as 2/5 of the ginned lint as the cost of ginning.

In a gin stand, round saws rotate at a high speed between parallel metal bands called ribs. When seedcotton is fed to the gin stand, the revolving saws grasp the seedcotton and draw it through the ribs. Seeds are bigger than the space between the ribs where the saws run, so they do not pass through the ribs, but fibers are drawn through the ribs easily. Seeds fall to the bottom of the gin stand, while lint is collected on the other side of the ribs. The remaining lint on the saws can be collected with the help of brushes or air.

Saw gins come in various sizes from as few as 5-10 saws per machine for ginning small samples from experimental plots, particularly breeding material, to over 150 saws per stand for commercial ginning. The diameter of the saws varies, depending on gin size. A gin stand having 150 saws can gin up to three tons of lint per hour.

A typical saw ginning system will comprise the following processes: module feeder, boll separator, airline cleaner, feed control, first tower drier, first extractor, second tower drier, cylinder cleaner, second extractor, extractor feeder, gin stand, first lint cleaner, second lint cleaner, humidifier, press, sample collection, wrapping and tying. Two more functions added to the ginning process are heating to adjust moisture level in the seedcotton and cleaning to get rid of trash. High moisture in seedcotton has a negative impact on the effectiveness of a gin's cleaning machinery.

Cotton fiber is hygroscopic. It can absorb and release moisture depending upon the relative humidity in the surrounding atmosphere. Ginning at high moisture levels produces low grades of cotton. Drying seedcotton to less than 6.5-8.0% moisture improves lint grades but reduces staple length and enhances short fiber content.

After ginning, moisture restoration to optimum levels reduces the force required to press a bale of cotton. Moisture can be restored by subjecting seedcotton to warm humid air or by spraying cotton with a fine mist of water. The mist method is usually used to restore moisture in lint just prior to pressing.

Ginning causes neppiness in cotton lint. Maintaining proper moisture and eliminating unnecessary processing steps during ginning reduces nep formation.

Cotton Cleaning

Seedcotton picked from the field contains trash from plant parts, i.e., leaves, leaf petiole, burs, branches and weeds. Trash must be removed to an optimal extent during ginning. Ginners tend to remove as much trash from cotton as possible, but pre-cleaning and post-cleaning, causes fiber quality damage. Thus, the ginning process is a compromise between separating lint from seed, getting rid of trash, maintaining fiber quality, and at the same time improving classer's grade. The addition of lint cleaners for removing trash at the gin increases the percentage of intermediate fibers and reduces the percentage of longer fibers. Lint cleaning is the last chance to improve the grade of cotton. However, excessive lint cleaning may unnecessarily remove fibers and result in weight loss.

Several kinds of seedcotton cleaning equipment are used to remove trash, some of which are cylinder cleaners, bur cleaners, stick machines and extractor feeders. Cylinder cleaners come in various sizes and inclinations and can be air-fed or gravity-fed. An extractor feeder is a system to uniformly feed seedcotton to the gin stand while performing cleaning as a secondary function.

Cleaners are used prior to the actual ginning operation (pre-cleaners) and after the ginning operation (post-cleaners), before lint is baled. The amount of trash in seedcotton determines how much cleaning is required, i.e., one or two pre-cleaners and post-cleaners.

Machine-picked cotton usually requires two seedcotton cleaners and two lint cleaners, compared to none or only one cleaning in the case of hand-picked cotton.

Lint cleaning removes small trash particles that cannot be removed during seedcotton cleaning. Heating and cleaning at the gin are used to optimize the cotton grade for maximum monetary returns to the farmer.

The cleanability of cotton can be improved by drying. However, drying enhances fiber breakage, thus reducing ginning out-turn and increasing short fiber content. The recommended moisture level in seedcotton for the least damage to fiber is 6.5-8.0%.

Cotton Baling

A bale is a compacted mass of staple fibers wrapped and ready to be shipped to a spinning mill. The weight of a bale varies among countries, from less than 50 kg to over 300 kg. Bales are packed in various sizes, weights and bale densities. Lint samples are usually taken at the time of baling cotton for classing.

Various kinds of material are used to wrap cotton bales. Polypropylene and light-weight jute cloths are the predominant materials used to wrap cotton bales, although cotton cloth is recommended. Polyethylene material is used to wrap bales in some countries. Wrapping material other than cotton may contaminate cotton.

Ties of round or flat metallic wire are used to hold bales together. Eight to ten ties are common.

Gin waste collected at various stages of cleaning has various uses. Depending upon the situation, early seedcotton waste can be fed to livestock or composted to be used as mulch. Lint cleaning waste can be processed to collect small fibers and motes for various uses.

Spinning and Weaving

Spinning is the process of making yarn from loose fibers. The most crucial part of spinning is the insertion of twist into a continuous strand of overlapping fibers to form a yarn. Twisting is preceded by many operations, such as carding, drawing, etc., which also form part of the spinning process.

Cotton was first spun by use of machinery in England in 1730. Developments in spinning machinery in 1730 and saw gins in 1793 paved the way to make cotton the most important natural fiber in the world.

SPINNING METHODS

Fiber bundles can be twisted in many ways. The three main technologies used on a commercial scale are ring spinning, rotor spinning and air jet spinning.

Ring spinning is the process of inserting twist by means of a rotating spindle. In ring spinning, twisting the yarn and winding it on a bobbin take place simultaneously and continuously. Ring spinning is a comparatively expensive process because of its slower speed; however, yarn quality is better. The additional processes (roving and winding) required in ring spinning make the process slower. Most of the yarn produced in the world is ring spun.

Rotor Spinning (Open-end Spinning) inserts twists by means of a rotating conical receptacle into which the fiber is admitted. In open-end spinning, air current and centrifugal force carry fibers to the perimeter of the rotor where they are evenly distributed in a small group. The tails of the fibers are twisted together by the spinning action of the rotor, and the yarn is continuously drawn from the center of the rotor. The process is very efficient and reduces the cost of spinning.

Open-end spinning eliminates the need for making a roving. At a speed of 60,000 revolutions per minute, the production rate of open-end rotors is 3-5 times higher than that of ring spinning. The yarn from open-end spinning is much more uniform compared to that from ring spinning. However, it is considerably weaker and has a harsher feel. Thus, low micronaire but stronger cottons are desirable for open-end spinning. Dust and trash that accumulate in rotor groves and interfere with spinning is the major problem of open-end spinning.

Air jet spinning (vortex) inserts twist by means of a rotating vortex of compressed air. The process is less expensive due to high speed. However, like rotor spinning, yarn strength is lower.

The *Compact spinning* system aims at improving yarn quality through narrowing the sliver leaving the drafting apparatus, before its twisting into yarn, and through liquidation of the twist triangle. The first spinning machine was built in 1995. The yarn produced in the compact system is smoother, has less hairiness and is usually stronger.

Important steps in the spinning process are opening and blending, cleaning, straightening or paralleling the fibers, sliver formation and twist insertion. Opening and blending is the first operation at the spinning mill. A bale is opened to make the lint fluffy, and bales are mixed or blended to ensure homogeneity. The main objective of cleaning at the spinning mill is to remove contaminants/extraneous matter from cotton lint.

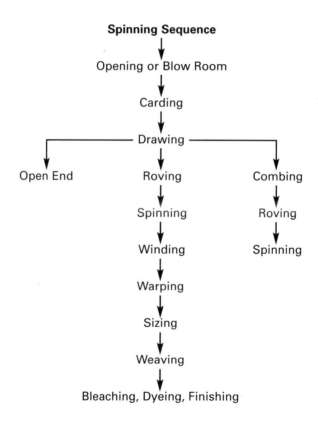

Spinning Sequence
↓
Opening or Blow Room
↓
Carding
↓
——— Drawing ———
↓
Open End   Roving   Combing
↓
Spinning   Roving
↓
Winding   Spinning
↓
Warping
↓
Sizing
↓
Weaving
↓
Bleaching, Dyeing, Finishing

Carding

The first mechanical action on individual fibers is to make them straight. Carding separates fibers from each other, aligns, parallels and condenses fibers into a single continuous strand of overlapping fibers called a "sliver." Impurities, including short fibers, are removed in carding.

Drafting

The drafting systems used in spinning mills are all roller drafting. The first pair of rollers rotates at a lower speed than the second pair of rollers. Drafting occurs when the first pair of rollers withdraws fibers as they are being delivered to the second pair of rollers rotating at a higher speed.

Combing

Combing is the process of removing short fibers and residual impurities from cotton that has already been carded. Combed yarn is superior in quality than carded yarn because it is stronger, more uniform and has fewer protruding fibers.

Noil. Short fibers obtained while processing cotton in a spinning mill are referred to as noil. Noils are mostly extracted during carding and combing operations. Noils contain spinnable and very short non-spinnable fibers.

Drawing

Drawing is the process of sliver drafting, doubling and redoubling. As a result of drawing the sliver becomes more uniform.

Slubbing and Roving Frames

Sliver needs to be condensed into a thinner strand before ring spinning. Slubbing and roving frames perform this function. The slubbing frame condenses the sliver to a thickness of almost one centimeter. The roving frame further condenses the sliver to a few millimeters.

Lap Formation

Fibers are straight in a sliver, but they are still in close association with one another. Lap formation is the process of spreading the sliver into a thin layer of fibers. The sliver lap is usually 25 centimeters wide. Fibers in the lap are easy to unwind and manipulate.

Floating Fibers

Fibers that in a drafting zone are not clamped by either of the pairs of rollers of the drafting system are floating. Floating fibers, due to not being held by any clamps, change their speed compared to other fibers and are thus responsible for producing less uniform yarn. Because all fibers are not of exactly equal length, the best drafting adjustment is a compromise between stretched fibers and floating fibers for each type of cotton being spun.

Yarn Count

The count of a yarn is a numerical expression of fineness. It is a yarn numbering system based on length and weight originally used for cotton yarns and now employed for most staple yarns. English counts are based on a unit length of 840 yards (768 meters), and the count of the yarn is equal to the number of 840-yard (768 meters) skeins required to weigh one pound. Under this system, referred to as indirect system, the higher the number, the finer the yarn.

In what is known as a direct yarn counting system, the yarn number or the count is the weight per unit length of the yarn. The direct system has two units: Tex (t)/ Decitex (dt) and Denier(d). The tex number is defined as the weight in grams of one thousand meters of yarn. The decitex number is defined as the weight in grams of ten thousand meters of yarn. The denier is defined as the weight in grams of nine thousand meters of the material.

The indirect system is more popular in the cotton spinning industry.

Yarn Strength

The strength of yarn depends mainly on the strength of constituent fibers and partly on the degree to which the fibers cling together and resist slipping over one another when the yarn is pulled lengthwise.

Below are a number of terms used in the processing of cotton into cloth:

Bobbin. A spool-like device upon which a filling yarn is wound for use in a shuttle in weaving. The bobbin set in the shuttle carries the yarn across the loom. With regard to the sewing machine, the bobbin is a metal receptacle that holds the under-thread beneath the metal plate below the needle.

Cone. With the help of winders, yarn is unreeled from bobbins and wound into a cone weighing about 2.25 kg. It is not economical to form a cone directly from spinning. Several bobbins will form a cone and the process is automated, including rejoining of broken ends.

Plying. Plying is the process of twisting two or more yarns together for specific end uses.

Wrapper: The wrapper is a machine that winds yarn from approximately 600 cones on to a spool-like beam for further processing.

Woven Fabric. Weaving is the art of interlacing yarns at right angles, much like making a basket.

Warp and Filling. In woven fabric, yarns run lengthwise and widthwise. The vertical or lengthwise yarn in the fabric is called warp and the horizontal or widthwise yarn is called filling.

Nonwoven Fabric. Nonwoven fabric is prepared directly from lint without the spinning process. Fibers are bonded together to form a fabric through mechanical entanglement or resin bonding. Other processes can also be used to produce a nonwoven fabric. Common forms of non-wovens are diapers, disposable wipes, and many kinds of paper including currency paper. Resin bonding or thermoplastic bonding is commonly used to prepare nonwoven fabric.

Knitted Fabric. Knitting is the art and science of making fabric by interlocking yarn around and through one another much like hand-knitting and crocheting. The most essential unit in a knitted fabric is the loop or stitch.

Bleaching. Bleaching is the process of removing natural and other types of impurities and blemishes from a fabric prior to dyeing and finishing. The most commonly used bleaches are chlorine, peroxide and reducing agents like sulphites.

Mercerization. Mercerizing is a finishing process used exclusively on cotton yarn and cloth consisting essentially of impregnating the material with cold, strong, sodium hydroxide (caustic soda) followed by washing. The treatment increases the strength and affinity for dyes and, if done under tension, greatly enhances the luster.

Sanforization. Sanforization is a treatment applied to fabric so that additional shrinkage will not exceed 1% in either direction.

Pilling. It is the formation of small fiber balls on the surface of cloth.

Count of Cloth. The number of ends and picks per inch in a woven fabric is called count of cloth. A cloth of 60x50 means there are 60 ends and 50 picks in one square inch of fabric.

Gray Cloth. A raw cloth that has not received any finishing after it is woven or knitted is gray cloth. Gray does not refer to the color of the cloth.

Textile. Traditionally, a textile is defined as a woven fabric made by interlacing yarns. The name derives from the Latin verb texere—which means to weave. But, in the broad market sense the word textile has come to include any yarn, fabric or cloth that has been made on a loom, knitting machine, or even needled, sprayed, spun bonded, or entangled non-wovens.

Abrasion Resistance. Abrasion resistance is the degree to which a fabric is able to withstand surface wear, rubbing, chafing and other frictional forces.

Washable. A cotton fabric that will not fade or shrink after washing or laundering.

Wrinkle free/Wash-and-wear. A garment that can be washed and used without ironing was first called drip-dry, but since all fabrics after all can be washed and drip-dried, the term was discarded. The second term used was wash and wear. Again, all fabric can be washed and worn so this was also discarded. The substitutes now are non-iron, easy-care or wrinkle-free.

Elasticity. Fabrics can be stretched and subjected to various types of distortions, but when allowed to relax they are expected to return to their original shape or size. Elasticity, or the ability of fabric or fiber to return to its original size, is only about 5% in cotton fiber or fabric, much lower than in most synthetic materials.

Fabric. An important branch of textile technology is the geometry of fabric structures. The appearance, handle, drape and general performance of fabric are dependent on the type of fiber used, yarn structure, weave and finish. One important dimension is yarn diameter, assuming that it is circular in cross section. The thickness or the fineness of yarn is determined in terms of linear density, meaning the fiber weight per unit length.

At the mill, short fiber content increases after opening and carding, but shows a significant decrease after the combing process.

The appearance of a fabric is largely dependent on yarn regularity and the frequency of neps, slubs and other thick and thin places.

Pricing and Price Risk Management

In addition to consolidating price information, the cotton futures market provides a means for all sectors of the cotton trade to manage or hedge their exposure to the risk of unexpected price fluctuations. Dramatic fluctuations of cotton prices may be attributable to a number of factors, ranging from weather changes in cotton producing regions to government policies. Cotton producers, ginners, merchants, and textile mills employ the cotton futures market to achieve price protection, reduce their effective purchasing costs or increase their ultimate selling price. By hedging the price of cotton they must buy and sell, they can avoid the potentially devastating effects of unexpected price fluctuations. This is possible because other market participants, professional traders and speculative investors are willing to assume the risk in return for the opportunity to profit should the market move in their favor. The following terms and definitions are used in the *cotton price risk management*.

An option is *In-the-Money* when an option's strike price is the same as the current trading price of the underlying futures contract.

Basis is the difference between the specific futures contract price and the cash price for cotton at a local delivery point. Normally, the futures price should be equal to the present cash price plus the cost of storage, insurance and interest charges necessary to carry the commodity to the delivery month of the contract. In addition, basis pricing also reflects the location (port of delivery) and the quality differential of the commodity.

Basis Risk is the risk associated with a widening or narrowing of the basis between the time a hedge is established and the time it is liquidated.

Buyer is a market participant who takes a long futures position or buys an option. An option buyer is also called a taker, holder or owner.

Call Option is a contract that gives the buyer/taker the right to buy the underlying futures contract at a stipulated price (strike price) at any time up to the expiration of an option, or to establish a long position in cotton futures. The buyer pays a premium to the seller/grantor for this contract. A call option is bought with the expectation of a rise in prices.

Cash-Settled is settling or meeting the obligations imposed by an expiring contract by means of cash payment, as opposed to physical delivery or receipt of a commodity. Cash payments are determined by value of contract, or market-to-market price. Most contract holders eliminate their positions through offsetting positions. Less than two percent of all futures contracts are held until expiration. If a futures contract holder chooses to hold the contract to expiration, a cash settlement may be made. A cash settlement essentially represents the difference between the contract value at the time of original purchase or sale and the contract value at expiration. However, the buyer and seller do not exchange the full value of the contract, but the difference in the contract value, compared with the original price.

Clearing House is the entity of the exchange responsible for recording transfers and settlements of exchange traded futures and options and the guarantor of all options. It is also charged with assuring the proper conduct of the exchange's delivery procedures and adequate financing and trading. When accuracy of all reported transactions has been verified, the clearinghouse then assumes the financial obligations to all traders. It becomes the buyer to every seller and the seller to every buyer. All contracts not offset by the end of trading are settled by delivery. A seller with open short contracts during the delivery period must either deliver the product or buy back the futures contract originally sold. Delivery can only be made from a location or a warehouse approved by the exchange. The buyer of a futures contract who has not liquidated his long position will receive and must accept delivery at some permitted location and must make payment in full for the product.

Closing-Out transaction is liquidation of an existing long or short futures or option position with an equal and opposite transaction; also called offset.

Collar is a compound option position consisting of a put and a call option in order to establish a fixed floor and ceiling. Collars are also known as fences, min-maxes or other similar names.

Commodity Futures Trading Commission (CFTC) is a U.S. federal regulatory agency established in 1974 to oversee futures trading and the operation of organized exchanges in the United States.

Cotlook Indexes (A and B) are published daily by a private company Cotlook Ltd. and are intended to represent the level of competitive offering prices for cotton delivered to North Europe on the international raw cotton market. The Cotlook A Index is an average of the cheapest five quotations from a current selection of the 16 Cotlook quotations for principal upland cottons traded internationally. The A Index was first introduced in 1966. The base quality of the A Index is middling 1-3/32" . The Cotlook B Index is an average of the cheapest 3 from a current selection of the 8 Cotlook quotations for coarse count cotton (such as strict low middling 1-1/16") used for spinning coarse count yarns. The B Index was first introduced in

1972. The geographical basis for the Cotlook quotations is North Europe, and the terms quoted are CIF (Cost, Insurance and Freight). See Cotlook Quotations.

Cotlook Quotations are estimated daily by the editorial staff of a private company Cotlook Ltd. based in Liverpool, and are intended to indicate the competitive level of offering prices. Prices are CIF North Europe, cash against documents on arrival of vessel, including profit and agent's commission. Cotlook collects information on offering prices from many sources and makes an assessment of prices with an unavoidable element of subjectivity. The Cotlook quotations are not contract prices, and traders are not guaranteed to conclude business at quoted levels. The Cotlook quotations are used to calculate A and B Indexes to represent the level of competitive offering prices on the international raw cotton market.

Cotlook Dual Index System is intended to provide two sets of competitive levels of offering prices for cotton to be shipped both during the current season and the next season. The nearby or current season's pair of A and B Indexes are for cotton to be shipped no later than August/September and forward season's pair of Indexes are for cotton to be shipped no earlier than October/November. Forward Indexes are usually introduced early in a calendar year as soon as next season's offers are available. See Cotlook Quotations and Indexes.

Cotlook Limited is a private company founded at the end of 1920's in Liverpool and is the publisher of Cotton Outlook. The company publishes cotton news and has been compiling and publishing the Cotlook A Index of raw cotton values and other quotes for cotton of different varieties since 1982. The A Index is used in many different ways by various countries and trading organizations. Cotlook Ltd. sells its cotton news and price services by subscription, including CIF Quotes, Cotlook Daily Internet Service, and Cotton Outlook Weekly Internet Service, Price Series and other information services. Cotlook Limited is a subsidiary company of The Outlook Group Limited. Other members of the group include Liverpool Cotton Services, which is a provider of cotton arbitration services.

Exercise is to elect to buy or sell, taking advantage of the right (but not the obligation) conferred by an option contract.

Exercise/Strike Price is the price, specified in the option contract, at which an options holder may buy or sell the underlying futures contract upon exercise of his option.

Expiration Date is the date on which an option contract expires; the last day an option can be exercised.

Floor Broker is a person who, in or surrounding any pit, ring, post or other place provided by a contract market for the meeting of persons similarly engaged, executes for another any orders for the purchase or sale of any commodity for future delivery and receives a prescribed fee or commission.

Floor Trader is an exchange member, also called a local, who usually executes his own trades by being personally present in the pit or place for futures and options trading.

Forward Contract is an agreement between buyer and seller for the sale of cotton of a specified quality and quantity for delivery at a specified future date. A price may be fixed in the contract; parties may agree to fix the price at any date in the future or upon delivery. Forward contracts usually are not traded on an organized exchange and most often result in the physical delivery of cotton. Forward contracts are also called cash contracts.

Fundamental analysis is an analysis concerned with economic factors comparing the relationship between the supply and demand for a given commodity.

Futures contract is a commitment to make or take delivery of a specified quantity and quality of cotton at an agreed price at a specified future date. Futures contracts are traded on an organized exchange. Futures contracts have standard delivery dates, trading units, terms and conditions established by the exchange. Futures contracts are sold and purchased through members of commodity futures exchanges. Qualified members may execute transactions for commission brokers, be retained by independent clients, or speculate for profit. Futures markets provide a number of advantages. They represent an efficient way of buying and selling commodities for future delivery, particularly seasonal farm crops. A futures contract also represents a valid agreement of purchase or sale of a physical commodity.

Futures hedge is a zero sum arrangement, when what is gained on one side of the market is forfeited on the other (futures or cash).

Hedging is taking a position in futures or options market offsetting a position held in the cash market to minimize the risk of financial loss from an adverse price change. The major function of a futures market is to provide facilities where commercial companies can hedge their price risks. The price to the public is reduced, because hedging enables processors to reduce operating costs. The ability to reduce risk by hedging makes the cost of financing lower. Basic hedging with futures contracts is achieved when a position in the futures market is approximately equal and opposite to the position held by the hedger in the cash or physical market. There is often a strong relationship between futures and cash markets. The quotation for each unit is expressed in the same way in both the cash and futures markets. The fact that traders know that deliveries can be made is the principal reason why futures prices preserve a continuing relationship with cash market prices. Since futures market prices typically move in tandem with the cash market over the course of time, tending to converge as contracts mature, a gain in the futures market will offset a loss in the cash market, or vice versa.

In-the-Money is an option with intrinsic value. For calls, the strike price must be below the current market price of the underlying futures contract. For puts, the strike price must exceed it.

Intrinsic value is a measure of the value of an option if exercised immediately and represents the profit differential between the option strike price and the current underlying futures market price.

Last Trading Day is a day on which trading ceases for the maturing (current) delivery month. The NYBOT rules specify the last trading day in cotton #2 futures contract as the 10th business day prior to last delivery date, which is the 7th to the last business day of the month.

Long - describes the position of someone who has bought cotton, futures contracts or options and has not yet offset that transaction with a sale or delivery of cotton. The opposite of Short.

Long Hedge is a purchase of a futures contract in anticipation of a cash market purchase. It is usually used to protect against a rise in the cash price.

Margin is a good faith deposit or collateral (usually a small percentage of the contract's full value) deposited by a client with his broker, or by a broker with the clearing house to ensure that market participants will meet their contractual obligations against open futures contracts. The margin is not a partial payment on a purchase. An initial or original margin is the total amount required by the broker per contract when a futures position is opened. Because the exchanges' clearinghouse guarantees contract performance to its members, the exchanges and the clearinghouse establish minimum margin levels for each market and periodically adjust them to reflect market activity, especially price volatility. Maintenance margin is a sum which must be maintained on deposit at all times. If a customer's equity in any futures position drops under the level because of adverse price action, the broker must issue a margin call to restore the customer's equity. See Variable limit margins.

Margin Call is a request from a brokerage firm to a customer to bring margin deposits up to minimum levels, or a request by the clearing house to a clearing member to bring clearing margins back to minimum levels required by the clearing house rules. Fluctuations in the futures market result in daily debits or credits in the margin account of a contract holder and cause margin calls.

Mark-to-Market is a daily cash flow system used by U.S. futures exchanges to maintain a minimum level of margin equity for a given futures or option contract position by calculating the gain or loss in each position resulting from changes in the price of the futures or option contracts at the end of each trading day.

Minimum Price Fluctuation is the smallest allowable increment of price movement in a given contract, also known as a tick.

Naked Position is an unprotected long or short position in a cash or futures market.

Naked Option is the sale of an option without holding an offsetting position in the underlying futures contract.

Nearby is the nearest listed-trading month of a futures market.

Notice days are when a notice of delivery shall be issued and tendered. According to the rules of the NYBOT it is the fifth business day prior to the day of delivery.

Offset is liquidating an existing long or short futures or option position with an equal and opposite transaction.

On call sales are a customary practice used in the physical cotton trade and are stipulated in the rules of the Liverpool Cotton Association, widely used in the international cotton trade. On call sales are agreements to sell cotton with the price to be fixed at a later time, based on a specified quotation by the seller or the buyer. If the final price is to be fixed based on a futures market quotation:

On *buyer's call*, the seller must fix the final price of cotton sold on call on the buyer's instructions, before the first notice day of the futures contract month. This must be done before the invoice is sent. If the buyer does not give instructions in the time agreed, the seller can fix the price when the market closes on the day before the first notice day of the futures contract month. The seller must notify the buyer about this immediately. If the parties agree that the price can be fixed after shipment, the cotton must be invoiced at a provisional price, and the difference will be paid by one of the parties based on the final price.

On *seller's call*, the roles of the buyer and the seller are reversed.

Open interest is the total number of futures or options contracts in one delivery month or one market that have not yet been offset by an opposite transaction nor fulfilled by delivery (one side only).

Open outcry is a public auction wherein verbal bids and offers are announced in trading rings at an exchange.

Option on a futures contract is an agreement between two parties that gives a buyer/holder the right, not the obligation, to buy or sell a specified quantity of futures contracts at a specified price (strike price) anytime on or before the expiration date regardless of the market price of that commodity. The buyer/holder pays a premium to the seller/writer for this option. There are no margin calls for purchases of options. An option contract has the same standardization of quality, quantity and location as a futures contract. There are two types of options: Call option and Put option. Options became an important hedging tool. Each option has a price paid by the buyer to the seller and determined in the marketplace. One of the largest advantages to buyers of cotton futures options is that the buyer can enjoy the price protection afforded by the futures market in the event of unfavorable price moves, while sharing the economic benefits of favorable price changes. Options' positions can be discharged by coming to expiration, usually the second Friday of the month preceding the underlying future expiration. Options can be liquidated by an offsetting transaction. Options are a flexible price risk management tool. Option buyers enjoy limited risk with unlimited profit potential. They can never lose more than the premium they paid. Users can fix prices at various levels, in effect using options as price insurance. Options make significant new pricing strategies available.

Out-of-the-Money is a call (put) option in which the strike price exceeds (is less than) the current market price of the underlying futures contract.

Position is an interest in the market in the form of open commitments, either long or short.

Premium is an amount paid by the buyer of an option to its seller. Premiums are determined in a trading pit and have two components: intrinsic value and time (extrinsic) value.

Price Limits are usually set by exchanges for daily price moves. The NYBOT sets price limits at 3 cents per pound, which in certain cases can be expanded to 4 cents per pound. There are no price limits on the expiring contract on or after its first notice day.

Put Option is a contract that gives the buyer/taker the right to sell the underlying futures contract at a stipulated price (strike price) at any time up to the expiration of an option, or to establish a short position in cotton futures. The buyer pays a premium to the seller/grantor for this contract. A put option is bought with the expectation of a decline in prices.

Seller/Writer/Grantor of an option is an individual who sells an option, establishing a short position.

Series of Options are options of the same type (either puts or calls, but not both), covering the same underlying futures contract, having the same strike price and expiration date.

Settlement Price is a daily price at which a clearing house clears all trades. It is based on the closing range of that day's trading and is the basis for both margin calls and (if applicable) the next day's price limits.

Short describes the position of someone who has sold cotton, futures contracts, or options and has not yet offset that transaction with a purchase or delivery of cotton. This is the opposite of Long.

Short Hedge is the sale of a futures contract in anticipation of a cash market sale. It is usually used to protect against a decline in the cash price.

Southern Mill Rules. With the organization of the American Cotton Shippers Association in 1924, the American Cotton Manufacturers Association agreed to jointly sponsor trading rules for the shipment of cotton to southeastern mills. The Southern Mill Rules were agreed to in 1925. Previously, trades were governed by the Carolina Mill Rules of 1915, or by individual mill terms. In the early 1900's, most cotton textile manufacturers were located in New England. Contracts between merchants and mills were made under the New England Mill Terms and quality arbitrations were handled by the New England Cotton Buyers Association located in Boston, Mass. With the movement of textile manufacturers to the southeastern states and the merger of the American Cotton Manufacturers Association with the northern Cotton Textile Institute in 1949, creating the American Textile Manufacturers Institute, Inc., the New England Mill Terms were rescinded in the 1950's. Revisions to the Southern Mill Rules to meet changed conditions of trade are considered yearly in a meeting between the Cotton Committee of ATMI, the Executive Committee of the Domestic Mills Committee of ACSA, and Amcot.

Amcot was founded in 1971 as an association of the four major U.S. cotton marketing firms. They are Calcot, Ltd., PCCA (Plains Cotton Cooperative Association), Staplcotn (Staple Cotton Cooperative Association) and SWIG (Southwestern Irrigated Cotton Growers).

American Textile Manufacturers Institute (ATMI) is the national trade association for manufacturers of textile mill products made in the United States. As such, ATMI's primary purpose is to provide its members with a forum to develop united approaches to indus-try-wide, national issues. ATMI's members produce approximately 80 percent of all textiles made in the United States, and process some 75 percent of all domestically grown cotton consumed annually in the United States.

Speculator is one who assumes the price risk that hedgers seek to avoid with a goal of receiving a profit resulting from price movements in the futures or options markets. Speculators provide liquidity and have the equity to absorb changes in price levels. Large quantities of risk capital are attracted to one location. Speculators help minimize price fluctuations by increasing the number of bids and offers in the market place. A publicly known value for a commodity is created and an alternative market is provided.

Spreads are simultaneous positions in the same contract but in different months or simultaneous positions in related contracts for the same or different months. There are two spread modes.

Carry Mode is the difference between the prices of two contract months and equals the cost of carrying cotton during the interval between the expiration dates for the two contracts.

Inverted Mode is where prices for nearby contract months are higher than prices of future months. This typically occurs in tight markets.

Strike/Exercise Price is the price, specified in the option contract, at which an options holder may buy or sell the underlying futures contract upon exercise of his option.

Swaps are cash-settled, bilateral agreements between two parties. The swap transaction does not entail physical delivery of cotton. It is not exchange-traded, but rather an over-the counter (OTC) instrument offered by brokers to customers to protect against adverse price movements, usually utilizing a non-New York quote, in order to avoid basis risk. There is a concrete buyer and a seller for each swap contract, and parties are matched with the help of a broker. The buyer of a fixed price swap for a forward month is due money from the seller should the reference price in the forward month be higher than the agreed swap price on a pre-determined quantity of cotton. Likewise, the buyer will owe money to the seller should the reference price in the forward month be lower than the agreed swap price. By entering into the swap transaction, parties can effectively lock in a fixed price by entering into the physical cotton market in the future to match their swap positions with a physical cotton position. This hedging strategy is similar to hedging with cotton futures. However, there is no basis risk for quality descriptions different from the New York futures contract. Most of the terms of the swap contract such as quantity, reference quote, settlement month or date are negotiable between the parties. As opposed to exchange traded contracts, cotton swaps transactions involve a counterparty credit risk, and it is customary for counterparties to offer and/or request letters of credit for a specified percentage of the total value of the swap contract.

Technical Analysis is concerned with price data rather than economic factors, with a presumption that current prices already reflect supply and demand relationships. Price projections are made based on certain chart patterns.

Time (extrinsic) value is the portion of a premium that exceeds the intrinsic value. The time value of an option reflects the probability that the option will move into-the-money. Therefore, the longer the time remaining until expiration of the option, the greater is its time value.

Variable Limit Margin is the performance deposit required whenever the daily trading limits on prices of cotton futures are raised in accordance with exchange rules. In periods of extreme price volatility, some exchanges permit trading at price levels that exceed regular daily limits. At such times, margins are also increased.

Variation Margin is a payment required upon margin call.

Volume of trading is the number of contracts traded during a specified period of time. (one side only). It may be quoted as the number of contracts traded or as the total of physical units, such as bales or tons.

Cotton Exchanges

In the history of the world cotton trade there have been many exchanges trading futures contracts. Trading volumes on most cotton futures exchanges declined drastically after World War II, and many exchanges ceased operations during the 1960s as a result of government policies in many of the cotton-producing countries. Currently there are only two exchanges trading cotton futures and options: *The New York Board of Trade (New York Cotton Exchange)* in New York, and the *Bolsa de Mercadorias and Futuros (BM&F)* in Sao Paulo, Brazil.

The Alexandria Cotton Futures Exchange was established in 1861 and was the first exchange formally established for cotton futures trading. The futures contract in Alexandria was 250 kentares (about 2,500 pounds). However, the Alexandria exchange suspended futures trading after 100 years because the Government of Egypt centralized cotton trading and had a negative view of speculation. Alexandria remains a major trading center for Egyptian cotton, and most Egyptian cotton trading companies and the Alexandria Cotton Exporters Association are based there.

The Liverpool Cotton Exchange formally introduced futures trading in 1882 under the aegis of the Liverpool Cotton Association. Initially, the exchange listed four contracts: an American contract, an "empire contract and miscellaneous growths" (including mostly Indian cotton), and two Egyptian contracts of different qualities. The size of the contract was 50,000 pounds, and the contract was traded up to 25 months forward. During the 1950s and 1960s, volumes traded at the exchange declined, and it terminated trading in 1964. Liverpool remains a center of cotton trade where major international cotton merchants and the Liverpool Cotton Association are based.

The New Orleans Cotton Exchange (NOCE) started trading cotton futures contracts in 1891. The NOCE traded two contracts (50,000 and 25,000 pounds) with characteristics similar to the New York cotton contract. U.S. government policy led to an accumulation of large cotton stocks held by the Commodity Credit Corporation (CCC) between the 1950s and 1970s. As a result, volumes traded in the cotton market declined. Operations at the New Orleans Cotton Exchange were terminated in 1964. Just as cotton trade between New York and Liverpool led to the development of cotton futures exchanges in the two cities, at about the same time the New Orleans Cotton Exchange and the Le Havre Cotton Exchange were a product of cotton trade between France and its former colony of Louisiana.

Le Havre was a major cotton trading center in France, initially trading mostly U.S. cotton and later cotton from French colonies in Africa. *The Le Havre Cotton Exchange* began trading Futures contracts in 1882. The contract size was 24,350 pounds with delivery months traded for up to a year forward. The Le Havre Cotton Exchange collaborated closely with the New Orleans Cotton Exchange. The decolonization of Africa and smaller volumes of trade caused the Exchange to suspend futures trading in 1965. The port of Le Havre remains a major cotton entry point in Europe.

Osaka is a major cotton trading center in Japan. *The Osaka Sampin Exchange* traded a cotton futures contract between 1910 and 1941, with some interruptions. The futures contract in Osaka was equal to 4800 pounds and was based on cotton of American origin for up to six months forward. The Japan Cotton Traders Association is based in Osaka.

Shanghai was one of the largest centers of cotton trade in China. *The Shanghai Cotton Exchange* was established in 1911 with the help of British and Japanese traders following the opening of futures trading in Osaka. The Shanghai Cotton Exchange was trading Chinese cotton, and the size of the contract changed over the years from 4,000 to 8,500 pounds. There were two delivery months a year traded at the exchange. The exchange closed in 1941 following the outbreak of the World War II. After the War, the government of China opposed attempts to reopen the exchange.

Futures contracts first started traded at the *Bremen Cotton Exchange* in 1914 for a brief period and were suspended because of World War I. Trading in cotton futures resumed in 1925. The exchange traded American cotton in units of 25,000 pounds. Trading stopped again in 1939 because of World War II and resumed in 1956. However, post World War II cotton futures trading in Bremen failed to attract an adequate number of speculators. In 1971, the Bremen Cotton Exchange suspended futures trading and now remains a center for spot cotton trade in Germany.

The Bombay Cotton Exchange started cotton futures trading in 1922 when the East India Cotton Association was given the authority to administer cotton futures trading under the Cotton Contracts Act. The size of the futures contract was 19,600 pounds. In 1952, under the Forward Contracts Regulation Act, the futures contract was transformed into a forward contract. In 1966, the Indian government prohibited trading in cotton futures altogether. In 1998, India reintroduced the cotton futures contract. However, the lack of speculation, and the predominant practice of physical delivery of cotton, makes this contract function like a forward cash contract, rather than futures trading.

The Chicago Board of Trade, one of the primary commodity exchanges in the USA, introduced a cotton futures contract in 1924. The contract characteristics were similar in all respects to the NYCE contract except for the size, which was 25,000 pounds. The Chicago contract failed to attract major volumes of trading due to competition from the NYCE and the NOCE. The Chicago contract was terminated in 1964.

The Karachi Cotton Exchange introduced a cotton futures contract in 1955. The contract was based on "Sind NT fine" quality and traded for January, March, May and July delivery months. Cotton futures trading in Karachi ended in 1971 as a result of the introduction by the government of the cotton price program. Under the cotton program the Government of Pakistan provided price guarantees to cotton farmers, fixed ginning charges, levied export duties, and ultimately decided how much cotton could be exported. Currently, the Government in Pakistan does not intervene in cotton marketing and trading, but the exchange still facilitates only a small number of spot transactions.

Following the close of the Liverpool and Le Havre exchanges, an attempt was made to introduce a cotton futures contract on *The London Commodity Exchange* in 1969 in order to service the European cotton trade. The London contract was set at 10,000 kilograms and traded in U.S. dollars with delivery points in Belgium and The Netherlands. However, as in the case with other European exchanges, the contract did not attract adequate interest from traders and speculators and was terminated in 1975.

The Hong Kong Commodity Exchange introduced a cotton futures contract in 1977 along with sugar, soybeans and gold contracts. The contract was for 50,000 pounds of American cotton, with a delivery point in Galveston, Texas and traded in U.S. dollars. The Hong Kong contract was based on the New York contract and directly competed with it. The volume of futures trading in Hong Kong rose from 1,151 contracts in 1977 to 14,630 contracts in 1980. However, in 1981 the contract was terminated as the exchange shifted its activities from commodities to financial products. In 1995 the name of the exchange changed to the Hong Kong Futures Exchange.

Recognizing the need for hedging instruments for non-U.S. cotton, the NYCE introduced a *World Cotton Futures Contract* in 1992. With a trading unit of 50,000 pounds, the contract was to be settled based on a consecutive 5-day average of the Cotlook A index. Despite high expectations and the trading platform of the NYCE, the world contract was traded sporadically during its two years of existence and failed to attract significant volumes. The major reasons for its failure were the lack of an equivalent spot market with well defined quality specifications and physical delivery locations, an element of subjectivity in the nature of the A Index, and currency exchange risks for cotton of undefined origin. The world contract was terminated in 1994.

China National Cotton Exchange (CNCE) was founded in 1999 by the All China Federation of Supply and Marketing Cooperatives, was approved by the State Council of China (Mainland) and it is a non-profit service organization. The exchange provides a platform for cotton forward and spot trading for Chinese cotton including electronically registered transactions. The bulk of cotton traded at the exchange originates from the government stock. The exchange is headquartered in Beijing the CNCE is involved in facilitating transaction, price discovery and communicating information. The CNCE has trading rules and provides contract forms. Its main services include transaction settlement, delivery, quality inspection, warehousing and transportation, information and consulting. Membership on the CNCE is open to companies registered in China (Mainland) and includes cotton dealers, textile mills, cotton traders and other affiliated services companies. Foreign-based companies could become members of the CNCE in the future. CNCE publishes daily transaction prices and indexes and information on current cotton supply and demand.

New York Board of Trade/New York Cotton Exchange

The New York Board of Trade (NYBOT) is the *parent company* of the New York Cotton Exchange (NYCE). In 1998 the NYCE joined with the Coffee, Sugar & Cocoa Exchange (CSCE) to form the NYBOT in a six-year merger process. As a result of the tragic events of September 11, 2001, the NYBOT location at 4 World Trade Center in downtown Manhattan was lost, and on September 17, 2001, the NYBOT restarted operations in a backup facility in Queens, New York. It is expected that the NYBOT will unite with the New York Mercantile Exchange (NYMEX) and the Commodities Exchange (COMEX) in a Manhattan building.

The New York Cotton Exchange was founded in 1870 by a group of cotton brokers and merchants and is the oldest commodity exchange in New York. Since its founding, the exchange has been an integral part of the cotton industry and today is the world's premier marketplace for cotton futures and options trading. In 1966 the Citrus Associates of the NYCE was formed as an affiliate, where frozen concentrated orange juice futures and options are traded. In 1984 the NYCE created the FINEX division and started trading a number of financial futures and options contracts.

The primary economic purpose of the NYBOT cotton futures market is to provide a forum for price discovery and a tool for price risk management. Cotton futures prices are established throughout the trading day by open outcry through the actions of many diverse market participants with a large number of competing buyers and sellers. Price quotes are transmitted worldwide. These prices reflect the latest information about supply and demand and are determined in a trading pit with the narrowest spread between bids and offers possible.

The clearinghouse of the NYCE (now the New York Clearing Corporation, NYCC) was created in 1915 removing counter party credit risk and providing a system of financial safeguards. In 1984 the NYCE instituted trading in options.

At the NYCE a standard cotton futures contract is traded, called the Cotton No. 2 futures contract. Changes to the terms of the contract have been introduced over time with the most recent changes being upgrades to fiber quality standards. The current Cotton No. 2 futures contract is for 50,000 pounds of Strict Low Middling grade with 1 1/16" staple length. The contract permits delivery only of white color cotton of grades from good middling to low middling and light spotted grades of good middling to middling. The minimum fiber strength requirement is 25 grams per tex, and micronaire readings of 3.5 to 4.9 are allowed, with no premiums or discounts. The Contract is traded for five delivery months: March, May, July, October and December. The nearest ten delivery months are available for trade. For example in April 2003 the latest delivery month available for trade is March 2005.

Currently the NYCE is open for daily trading of the No. 2 contract and cotton options between 12:15 PM and 3:00 PM New York time on weekdays (except holidays). In 2002 the volume of Cotton No. 2 contracts traded at the NYCE was 2.3 million contracts, and 1.2 million cotton options were traded.

Bolsa de Mercadorias & Futuros

The Bolsa de Mercadorias & Futuros (BM&F) in Sao Paulo, Brazil launched a new cotton futures contract on November 22, 1996, desiring to have a regional hedging instrument to reflect the fundamentals of the Southern Hemisphere market, which are different from the Northern Hemisphere season, reflected by the New York futures market. The contract is elaborately designed and is housed at the exchange, which has accumulated significant experience in successfully operating a number of futures contracts and financial instruments. The BM&F futures contract was tailored in many respects similar to the NYCE futures contract. It is based on cotton produced in the states of Sao Paulo and Parana, Type Six, with a minimum fiber length of 1-1/16 inches. Prices are quoted in cents per pound and the Exchange establishes daily price fluctuation limits, except for the nearest month. Contract months are March, April, May, July, October and December. The BM&F contract size is just 10,000 pounds, or one-fifth of the New York cotton futures contract. The contract size was set smaller than New York to attract smaller Brazilian producers and speculators to the market. The delivery point is the city of Sao Paulo, State of Sao Paulo, and there are eight locations listed in the contract in the states of Sao Paulo and Parana, where licensed warehouses are located. In November 1999 the Brazilian Central Bank issued an authorization for foreign investors to operate in the futures market, and contracts could be settled in US dollars guaranteeing income tax exemptions for these operations. Another important measure introduced in 1999 was the inclusion in the contract of a stipulation on the physical delivery of cotton.

The contract stipulates elaborate rules for daily settlement of accounts and delivery procedures, which can be replaced by cash settlement in US dollars. The BM&F provides arbitration services. Margin requirements are fixed by the Exchange per contract and are due on the business day following each trading day. The positions outstanding in the nearby month are subject to double margins. Significantly, hedgers are granted a 20% discount on the initial margin. Assets eligible to meet margin requirements include cash, gold, and, at the Exchange's discretion, government bonds, private securities, letters of credit, insurance policies, stocks, and shares of equity funds.

During the first year of existence, the BM&F cotton futures contract was regularly traded on a daily basis and has demonstrated a significant amount of volatility, an essential element that should make the futures contract attractive to speculators. However, trading volume remained relatively small and averaged just 56 contracts a day, an equivalent of about 1,200 bales of cotton. A total of 13,689 cotton contracts were traded at the exchange in 1997. In 1998 the number of cotton contracts traded rose to 17,007 with an average daily at 70. In 1999 volumes of cotton futures traded at the exchange declined to 5,115, averaging 21 contracts a day. The trading volume continued to slide falling to 306 contracts in 2000; 15 contracts in 2001; and 75 contracts in 2002. The contract continues to be listed by the exchange, as of 2003, however, trade volumes remain insignificant.

Obviously, the BM&F cotton futures contract has not attracted significant interest and has not become a popular instrument for hedging and speculation in Sao Paulo. Traders initially expressed an interest in using the Sao Paulo contract, but they do not want to risk exposure while volumes are still low, preferring to use New York for their hedging needs. It seems that local speculators have not been attracted to the cotton futures contract in Sao Paulo and prefer to speculate in stocks, public debt and other well established futures markets. The liquidity of the market can not grow without speculators. The BM&F has a representative office in New York which provides information on the activities of the exchange.

International Cotton Organizations and Associations

Participants in international trade in many countries are organized into national cotton associations, many with an international character. The membership of some associations includes both domestic and foreign organizations and individuals. Cotton associations serve the interests of cotton producers, buyers, sellers and consumers by providing trading rules and mechanisms to resolve trade disputes and serving as arbitration authorities. Cotton associations promote and facilitate cotton trade in a fair and orderly fashion to the benefit of a sound world cotton economy. In addition to maintaining trading rules, cotton associations provide other important services, such as technical and quality arbitration, traditional and HVI classing, forums for international conferences and discussions of cotton affairs, training seminars around the world, market information and statistics.

Cotton associations differ in organization and representation, some serving mostly exporters or producers and others, importers and spinners. A number of the associations, such as ACSA, play an important role in formulating national cotton policies, which have a significant impact on international trade in cotton. Information, statistics and training provided by the associations is a valuable contribution to a better understanding of the complexities of the market, which many participants of the cotton trade can benefit from.

Fourteen of the largest cotton associations with similar objectives comprise the *Committee for International Cooperation between Cotton Associations (CICCA)*. Each of the CICCA member associations acts independently, but uses CICCA as a forum for discussion and collective action when appropriate. CICCA promotes trading rules and arbitration practices of its member associations and stands for the concept of sanctity of contracts and good trading practices. CICCA objectives include assistance in ensuring that dispute resolution procedures are adhered to and any consequential awards upheld. CICCA circulates to member-organizations a consolidated list of firms reported to have failed to properly comply with valid arbitration awards made by member-organizations. CICCA publishes a directory of all firms affiliated with its member associations. Membership in the 14 CICCA-member associations accounts for more than one thousand firms associated with the cotton industry. Members of these associations handle the bulk of world cotton trade.

The European Cotton Confederation (ECC)

The ECC was founded in 2001 as a result of efforts under the European Contact Group to harmonize cotton-trading rules for cotton trade in Europe. The purpose of the ECC is to produce a harmonized European contract accepted by all of the European cotton associations. The Liverpool Cotton Association, the Gdynia Cotton Association, the Bremen Cotton Exchange, the Associazione Tessile Italiana, Centro Algodonero Nacional in Spain, Association Française Cotonnière and the Belgian Cotton Association joined the confederation. It is expected that the European Contract could be launched in 2004.

African Cotton Association

African Cotton Association (ACA) was founded in 2002 and is composed of 27 private and public organizations from 11 countries in Western and Central Asia and is headquartered in Burkina Faso with a representative office in Paris. The Association was founded out of concern shared by many of the producing countries about government policies in some of the major producing countries subsidizing cotton production and leading to market distortions. The Association plans to work with other international associations on upholding fair cotton trading rules and the sanctity of contracts. Informational and educational work will play an important part in the association's activity.

The Alexandria Cotton Exporters Association

Cotton traders founded the Alexandria Cotton Exporters Association (ALCOTEXA) in 1932. As a non-profit and non-trading organization, ALCOTEXA deals with issues related only to Egyptian cotton. Non-members of ALCOTEXA are not allowed to trade cotton in Egypt. Membership in ALCOTEXA includes cotton trading and ginning companies. All exports of cotton from Egypt are subject to the terms of the Egyptian Contract. The Association has a Board, Management Committee, Expert Cotton Classers and Government Sworn Experts on the Arbitration and Appeal Boards. All export sales from Egypt are based on Egyptian Testing House Terms, and arbitration is provided in Alexandria. Major functions of ALCOTEXA as a regulative authority include formulation of the export policy and setting sale prices (indicative or minimum). The Association revises export prices weekly. The association conducts contract registrations. Twice a year the ALCOTEXA publishes The Egyptian Cotton Gazette, which contains a variety of statistics on Egyptian cotton, articles and data on trade, policy and technical issues. ALCOTEXA is a member of CICCA.

The American Cotton Shippers Association

The American Cotton Shippers Association (ACSA) is the national trade association in the USA of cotton merchants, cotton shippers and exporters of raw cotton, primary buyers, mill service agents, and of firms allied with these services. Its membership is composed of four Federated Associations: Atlantic Cotton Association; Southern Cotton Association; Texas Cotton Association and Western Cotton Shippers Association. ACSA has about 150 member firms, which handle an estimated 80% of the cotton sold to domestic mills in the USA. Many of the ACSA members are also members of other world cotton associations, including the LCA. ACSA firms account for the bulk of international cotton trade handled by members of those Associations.

ACSA was organized in 1924 and joined with the American Cotton Manufactures Association to formulate the Southern Mill Rules. ACSA, like the LCA, actively promotes fair trading practices, sanctity of contracts and requires strict adherence to contractual obligations and arbitration awards. ACSA provides an environment of fair trade through established rules and industry practices that encourage favorable resolution of all disputes. One of the objectives of ACSA is to educate producers and mill customers on the prudent and appropriate use of available risk management instruments that will enable them to maximize their profit potential. In conjunction with the American Textile Manufacturers Institute, ACSA maintains the Cotton States Arbitration Board in Memphis and has its own representatives on the appeal boards of cotton associations in Germany, France, Belgium, United Kingdom, Spain and Japan and has arrangements with the Bombay Appeal Board to be represented. ACSA is actively involved in evolving regulations, rules and arbitration procedures, governing cotton trade with importers of US cotton, dealing with cotton associations and exchanges.

ACSA actively promotes increased cotton use in the USA and throughout the world. ACSA is very active in international cotton affairs and was one of the initiators of the Committee for International Cooperation between Cotton Associations (CICCA). It represents the US cotton trade in various international forums, such as ICAC and ITMF.

One of the most important functions of ACSA is to represent the interests of the US cotton merchandising industry before the US government and Congress, advising government agencies of the industry's views on existing or proposed regulations and national laws. ACSA collaborates with producers, processors and users of cotton in formulating farm programs and marketing provisions affecting the commodity. It cooperates with other sectors of the cotton industry on issues related to cotton trade establishing specialized committees on quality and standards, gins, compresses and warehouses, futures contracts, domestic mills and others. Policy issues of ACSA are discussed at its annual conventions attended usually by several hundred participants.

In 1995, ACSA and Rhodes University in Memphis, Tennessee entered into an educational partnership designed for those entering the cotton industry or desiring to expand their knowledge of the trade in cotton. A comprehensive nine-week, residential, course of study is offered on all aspects of cotton production and marketing including U.S. and world production trends, classification and fiber properties, processing, merchandising, shipping, and manufacturing, with an emphasis on trade terms, domestic and international trade rules, ethics, forward and spot contracts, and the appropriate use of futures and options contracts.

Association Française Cotonnière

Association Française Cotonnière (AFCOT) is more than 100 years old and has about eighty members, including firms based in France and other countries. Membership includes cotton merchants, agents, shippers, controllers, transport organizations, ports, banks and spinners. AFCOT is ruled by a board of directors composed of members, usually merchants and controllers. AFCOT has several committees, including the Advisory Committee for Arbitration and Supervision of the Types, and the Committee on Value Differences.

AFCOT publishes Le Havre General Rules, which regulate contracts for the sale of cotton and arbitration. According to a 2003 estimate, up to 100,000 tons of cotton are traded annually Europe under AFCOT Rules. AFCOT has a laboratory which is equipped for fiber testing. The Association issues a news bulletin to its members with trade statistics and other cotton related data. The annual dinner of AFCOT is attended by hundreds of cotton representatives from France and abroad. AFCOT is a member of CICCA.

The Associazione Tessile Italiana

The Associazione Tessile Italiana dates back to 1883 and has close to three hundred member firms operating in raw cotton marketing, cotton and linen spinning, weaving and finishing industries in Italy. It is estimated that members of the Association account for 70% of the volume handled by the represented industries. The Raw Cotton and other Fibers Section of the Association focuses on cotton issues. The Raw Cotton Arbitration Chamber operates within the Association and serves to resolve disputes arising from cotton contracts based on the trading rules. The association has a technical laboratory equipped with modern instruments used for cotton fiber tests and research. Substantial efforts are devoted by the Association to the promotion of cotton textile products at fashion shows.

The Association represents the industry on issues related to international trade, trade duties and quotas, currency and customs regulations. It provides training and consulting services to its members, including financial analysis and market research, insurance and currency markets analysis. The Association is a major source of economic and statistical data produced by its Economic Research and Statistics Bureau. The Association has a variety of publications on cotton and linen technical and economic issues. The two-volume statistical yearbook contains national and international statistics on production, trade, stocks, consumption of cotton and other fibers. There are monthly statistical publications with data on trade, employment and output in the spinning sector.

One of the major functions of the Association is to negotiate the national collective labor contract for the industry. It maintains contacts and negotiates with Italian government authorities and trade unions in the industry. The Association provides specialized consulting services on direct and indirect taxation and fiscal issues. The association is a member of CICCA.

Australian Cotton Shippers Association

The Australian Cotton Shippers Association was established in 1984. It is comprised of all major merchants in Australia. The association trading rules serve to achieve the major objectives of the association, including preserving the sanctity of contracts, the integrity of the Australian trading industry and to facilitate compliance with contractual obligations and adherence to arbitration awards. The association promotes the interests of the members in overseas markets and is an active participant in CJCCA.

The Belgian Cotton Association

The Belgian Cotton Association is composed of about forty Belgium based merchants, brokers, spinners and controllers, and foreign firms registered as associate members, including merchants from the USA and Switzerland. Major functions of the Association include maintenance of trading rules and arbitration. The Association has an Executive Committee of the Chamber of Arbitration, which issues value differences for cotton of different origins. Twelve arbitrators serve in Quality and Commercial Arbitrations and Appeals. The annual report of the Association includes data on Belgian and international cotton markets. International cotton merchants and representatives from Belgium and abroad attend the annual dinner of the Belgian cotton Association. The association is a member of CICCA.

The Bolsa de Mercadorias & Futuros in Sao Paulo, Brazil

The Bolsa de Mercadorias & Futuros in Sao Paulo (BM&F), Brazil, was founded in 1917 and is an exchange where gold, currencies, and a number of agricultural commodities, including cotton, are traded in futures and cash markets. As far as cotton is concerned the Exchange has the elements of a cotton association of traders and as such is a member of CICCA. The Exchange plays an important role in regulating domestic trade, exports and imports of cotton in Brazil. The Exchange provides classification services and establishes standards for Brazilian cotton. The Exchange settles disputes between traders and provides arbitration. A special committee at the Exchange fixes value differences for different grades in relation to Type 6. A substantial part of all contracts traded in Brazil is made subject to the rules of the BM&F and is registered with the exchange. The Exchange collects and publishes statistics related to cotton and is active internationally in promoting the sanctity of contracts and fair trading practices.

The Bremen Cotton Exchange

The Bremen Cotton Exchange was founded in 1872 and now represents more than 200 merchants and users of cotton. The Bremen Cotton Exchange provides international trading rules, technical and quality arbitration, traditional and HVI classing. The Exchange conducts fiber testing and research and expert surveys. Trading rules of the Exchange regulate trade in raw cotton, linters, cotton and man-made fiber wastes and provide a basis for quality and technical arbitration and effective means for the settlement of disputes. Bremen cotton trading rules are used most widely in Germany, Switzerland and Austria.

Bi-annual International Cotton Conferences held by the Bremen Cotton Exchange deal mostly with technical issues and are attended by a large number of participants. The Exchange organizes seminars where participants receive training in cotton classing and other technical subjects. In 1969, the Bremen Fiber Institute was founded, which works as a laboratory for the Exchange. The Institute is equipped with an HVI test line and modern spinning and weaving equipment. The Institute's research is focused on cotton fiber properties for processing. The exchange is a member of CICCA.

The Bremen Cotton Exchange provides statistics and information on the domestic and international cotton market, technical issues, value differences, freight and insurance. The Exchange issues an annual report and the Bremen Cotton Report every two weeks.

Centro Algodonero Nacional in Spain

Centro Algodonero Nacional (CAN) in Spain was founded in 1903 in Barcelona. It represents all the sectors related to the marketing of raw cotton and its sub-products. The major objectives of the Centro are to create fair trading conditions and to promote the sanctity of contracts. The rules of the Centro are internationally known and recognized as the Barcelona Contract. The Centro has the capacity for quality arbitration and appeals. It has a laboratory, which can perform fiber testing. Membership of the Centro includes more than a hundred individual members, cotton firms and associate members. Member firms and about 30 individual members operate as cotton merchants, agents, or brokers. It is estimated that most of the sales of cotton in Spain are made subject to Barcelona rules. The Centro provides its members a variety of services, disseminates cotton information and statistics, and is a member of CICCA.

The East India Cotton Association

The East India Cotton Association (EICA) was established in 1922 and has about 400 members including buyers, sellers, brokers, exporters, importers and other participants in the cotton market. By-laws of the EICA provide trading rules for spot and forward sales of cotton. The Association is managed by the Board of Directors through various sub-committees. There are 18 regional associations and 10 marketing societies registered under it. The Rules of the EICA provide mechanisms for arbitration and settlement of disputes. The EICA has a panel of Sworn Surveyors, an Umpire and a provision for appeal.

One of the major functions of the Association is to prepare and maintain grade and staple standards of all varieties grown in India. The Association has a laboratory for fiber quality evaluation and conducts HVI cotton fiber testing. The Daily Rates Committee fixes and releases daily prices for various descriptions and staples and grades. The EICA publishes other market data in its weekly bulletin. The bulk of cotton traded in India is regulated by the rules of the Non-Transferable Specific Delivery Contract of EICA. The association is a member of CICCA.

The Gdynia Cotton Association

The Gdynia Cotton Association (GCA) was founded in 1935 and is composed of over one hundred member-companies from 15 countries. Membership of the GCA includes cotton textile mills, cotton merchants and research institutions. The GCA By-laws and Rules are used as the basis for international cotton contracts and stipulate in detail cotton arbitration and testing procedures. The major objectives of the Gdynia Cotton Association include settlement of quality and technical disputes arising from cotton trade by the Court of Arbitration at the GCA, laboratory quality testing and representation of the member's interests before government authorities and international organizations. The GCA provides cotton classification courses for cotton classers in Polish, English and Russian languages and publishes value differences tables and a directory of member firms. The Gdynia Cotton Association organizes international discussions at the bi-annual International Cotton Conferences in Gdynia. The Association collects statistics on cotton imports and consumption in Poland, which it publishes annually. The GCA is a member of CICCA.

Izmir Mercantile Exchange

The Izmir Mercantile Exchange in Turkey was founded in 1881. Like the BM&F in Sao Paulo, the Izmir exchange functions as a trading platform for spot sales of cotton and as an association of cotton traders. The exchange serves as a price discovery instrument for spot sales of cotton and has been engaged in designing a cotton futures contract for potential introduction on the exchange-trading floor. The exchange maintains trading rules, provides information and statistics, and is a member of CICCA.

The Japan Cotton Traders Association

The Japan Cotton Traders Association was founded during the 1950s and is composed of about 80 Japanese cotton importers, domestic raw cotton traders and firms engaged in related businesses, such as shipping agents, transportation and warehousing, banks and insurance. Members of the Association handle the bulk of cotton imports in Japan and imports by Japanese owned spinning mills in other Asian countries. The major objective of the association is to strive for the sound development of cotton import and domestic trade, trying to improve the basic terms and conditions for trade. The association is entitled to settle any claim or dispute, which may arise in connection with the import and domestic trade of cotton. JCTA makes recommendations to the government and its agencies and cooperates with other international associations and organizations on issues related to cotton trade. JCTA conducts research and collects statistics related to cotton and issues a number of publications, including a statistical yearbook. The association is a member of CICCA.

The Karachi Cotton Association

The Karachi Cotton Association (KCA) was established in 1933 to regulate and facilitate domestic and export trade in cotton. It has about 250 members, including cotton growers, ginners, textile mills, exporters, commission houses and others. The KCA is ruled by a 21-member board of directors, of which 17 are elected annually from the membership of the KCA and four are nominated by the Government and represent the Ministries of Agriculture, Commerce, Finance and Industry. The Rates Committee of the KCA, appointed by the Board, establishes daily spot rates based on cotton transactions throughout the country. By-Laws and Rules of the KCA regulate cotton trade and provide arbitration of disputes between parties.

The KCA provides traders with contract forms and adopts standards for cotton. The Association issues a Daily Cotton Market Report, containing information on spot prices and other statistics related to cotton. The KCA advises the government on various aspects of cotton policy and maintains liaison with ginners and textile mills. The Association founded the KCA Institute of Cotton Grading and Classing, which provides training to representatives of the cotton industry.

The Liverpool Cotton Association

The origins of the LCA date back to 1841 when cotton brokers in Liverpool formed an association and drew up a set of trading rules. In 1882, merchants joined brokers and formed a new association named The Liverpool Cotton Association. The membership of the LCA includes buyers and sellers of cotton, international merchants, government marketing organizations, spinners, banks, cotton controllers and others involved in the cotton business.

One of the major functions of the LCA, as well as other associations, is to reduce risks involved in international cotton trade and to provide an effective mechanism for settlement of disputes arising between parties involved in trade and to uphold equitable trading practices and the sanctity of contracts worldwide. The Association provides a set of by-laws and Rules, which are widely accepted and cover all aspects of international trade. Membership in the LCA is in excess of 300 registered firms in over 60 countries worldwide. It is estimated that over 60% of the world's cotton trade is bought and sold under the Liverpool Cotton Association Bylaws and Rules. The LCA has the largest share of registered firms, which are based overseas, compared with the other cotton associations. In fact it is the only association with a majority of member firms based overseas. About 600 official copies of the Rule Book are in use around the world.

Contracts made subject to Liverpool Rules are dependent upon Liverpool Arbitration in the event of a dispute between parties. The LCA provides a well-established two-tier arbitration system for both quality and technical (non-quality) disputes. Contracts written under LCA rules are subject to the Laws of England; however arbitration awards can be legally enforced in most cotton trading countries. In the event a firm refuses to abide by arbitration or appeals a decision, the firm is included on a default list, which is distributed among all members of the LCA and 13 other principal cotton associations world-wide, and may be suspended from registration with the LCA. The default list is reviewed annually.

One of the most important functions of the LCA is the training it provides on aspects of international trade in cotton. The LCA holds an annual marketing seminar in Liverpool and has also conducted tens of seminars in different countries. The seminars are devoted to the principles of contracting, arbitration, futures and options trading, banking and other aspects of international cotton trade.

The LCA also serves as a forum for international discussions of world cotton affairs and cooperates closely with other cotton associations and international organizations, such as ICAC, ITMF and CICCA. The LCA annual dinner is one of the major world cotton events and is usually attended by hundreds of members and guests of the Association.

The LCA provides cotton laboratory testing, including HVI testing. The LCA publishes a bimonthly Value Differences Circular, which contains quote differences for grade and staple for various growths, applicable in case of Quality Arbitration and Appeal. The LCA also issues a Directory of Membership, Contract and Arbitration Award forms and other publications.

Russian Cotton Association

The Russian Cotton Association was established in 2000 by leading Russian cotton trading companies and was joined by other cotton and textile related organizations as well as by a number of foreign merchants. The association represents the industry in government organizations and aims at establishing fair trading rules and the principles of sanctity of contracts in the Russian cotton market. It has created a volunteer dispute resolution court comprised of the members of trade organizations as a first step in establishing arbitration procedures. The Association provides a forum for discussion of issues of importance for the cotton merchandising industry and organizes international conferences twice a year during national textile fairs in Moscow.

ICAC

The International Cotton Advisory Committee (ICAC) is an association of governments having an interest in the production, export, import, and consumption of cotton. It is an organization designed to promote cooperation in cotton affairs, particularly those of international scope and significance. It affords its members a continuous understanding of the world cotton situation and provides a forum for international consultation and discussion. The Committee has consultative status with the United Nations and its specialized agencies and cooperates closely with other international organizations in matters of common interest.

The committee cooperates with the Common Fund for Commodities on cotton projects. The Common Fund for Commodities (CFC) is an international intergovernmental organization affiliated with the UN system and based in Amsterdam, The Netherlands. Ten cotton projects involving approximately $20 million in funding from the CFC have been approved since the ICAC was recognized as the international commodity body for cotton in the early 1990s.

The functions of the International Cotton Advisory Committee are defined in the Rules and Regulations, copies of which are available on request. These are:

To observe and keep in close touch with developments affecting the world cotton situation.

To collect and disseminate complete, authentic, and timely statistics on world cotton production, trade, consumption, stocks and prices.

To suggest, as and when advisable, to the governments represented, any measures the Advisory Committee considers suitable and practicable for the furtherance of international collaboration with due regard to maintaining and developing a sound world cotton economy.

To be the forum for international discussions on matters related to cotton prices.

The Committee is the outgrowth of an International Cotton Meeting held in Washington, D.C., in September 1939. At that time, world stocks of cotton had reached nearly 25 million bales, of which over half were located in the United States. The following ten producing countries were convened: Brazil, British cotton exporting colonies, Egypt, French cotton exporting colonies, India, Mexico, Peru, the Sudan, the USSR, and the United States, to discuss problems of over-production, rising stocks and falling prices. The principal objective was to take concerted international action to avoid chaotic developments in the world cotton economy.

The ICAC met for the first time in Washington, D.C. in April 1940. At first, membership was limited to cotton producing countries. After the fourth meeting, an invitation to join the Committee was extended to "all other United and Associated Nations substantially interested in the production, export or import of cotton." At the fifth Meeting in May 1946, it was decided to establish a Secretariat and a governing Executive Committee consisting of representatives from six cotton producing- and six cotton-consuming countries. A beginning was also made with the statistical and economic information program, which became an intrinsic part of the Committee's work. Subsequently, in 1948, it was agreed to replace the original Executive Committee with a Standing Committee in which all countries would have a voice.

Membership

Membership in ICAC is open to all members of the United Nations or of the Food and Agriculture Organization of the United Nations, expressing an interest in cotton. Any other government expressing an interest in cotton shall be eligible to apply for membership. The annual budget of the Committee is supported by assessments to member governments, subscriptions to publications and participant fees at meetings. Forty percent of the total assessment is shared equally among member countries. The remaining sixty percent is allocated based on the average trade in raw cotton in the four most recent cotton seasons. There are more than forty member-countries in the ICAC divided almost equally among cotton producing and consuming countries.

Plenary Meetings

At the invitation of a member government, plenary meetings of the Advisory Committee are held each year. Meetings alternate as much as possible between cotton exporting and importing countries. A provisional agenda and time schedule for each plenary meeting is prepared by the Standing Committee. Provision is made for the exchange of information on the cotton situation in various countries and for discussions on international policy matters. Since the mid-1960s, technical seminars on subjects of interest to research workers have been held as a part of the plenary meeting. The Secretariat of the Committee publishes the formal proceedings of each plenary meeting.

The official languages of the ICAC are Arabic, English, French, Russian and Spanish. Full simultaneous interpretation is provided at plenary meetings.

Standing Committee

A Standing Committee, consisting of representatives of all member governments, gives continuity to the activities of the International Cotton Advisory Committee between plenary meetings. The Standing Committee convenes in Washington. Attention is given to the current world cotton situation, cotton policy matters, and also to assignments from the plenary meetings.

The Secretariat

The Secretariat of the Committee is located in Washington, D.C., and is composed of an international staff headed by an Executive Director whose appointment and contract of employment is determined by a plenary meeting. The Secretariat assists the Committee in carrying out its work program by developing and publishing statistics and analyses of the world cotton situation; by carrying out a program of work to disseminate information on cotton production research; by cooperating with other organizations to promote a sound world cotton economy; and by facilitating discussions on matters related to cotton prices.

Coordinating Agencies

Each member government is required to establish a "permanent national coordinating agency" to provide the Secretariat with statistics on the cotton situation and to distribute publications and reports received from the Secretariat. Under this cooperative arrangement, there has been continued improvement in the quantity and quality of statistics on cotton available on a world basis. Coordinating agencies are frequently called upon to supply information requested by special inquiries from the Committee. Their cooperation has made possible work surveys on various subjects of importance to member governments.

Publications

Regular publications of the ICAC Secretariat include: *Cotton This Week, Cotton This Month, Commitments This Month, Extra-Fine Cotton This Month, COTTON: Review of the World Situation, The ICAC Recorder, COTTON: World Statistics, World Textile Demand, World Cotton Trade, The Outlook for Cotton Supply, Production and Trade Policies Affecting the Cotton Industry, Agrochemicals Used on Cotton, Bale Survey, Classing and Grading of Cotton, Current Research Projects in Cotton, Growing Organic Cotton, Insecticide Resistance and its Management in Cotton Insects, Survey of the Cost of Production of Raw Cotton, Survey of Cotton Production Practices, The World Market: Projection to 2005, Proceedings and Statements of the Plenary Meeting* and ICAC Documents on CD-ROM.

Trade and Shipping

International cotton merchants play a leading role in cotton trade and are engaged in the business of buying cotton from farmers or country merchants, selling to textile mills, both domestic and foreign, and arranging shipment to destination. Merchants provide the services of buying when farmers want to sell, selling when mills want to buy, aggregating cotton supplies to provide lots desired by mills, hedging price risk and arranging transportation. Some agencies specialize in the implementation of their government's decisions by executing purchases and sales through other international merchants. Merchants are vertically integrated in many countries, supplying credit, inputs, extension services, information and market news to producers, ginning cotton and marketing both seed and lint. Merchants commonly trade cotton among themselves to acquire particular types for individual customers.

The Secretariat of the ICAC has studied the structure of world trade in cotton since 1994 and has compiled a list of cotton-trading companies active in international trade. The list of cotton trading companies consists of about 500 firms engaged, at least in part, in international trade in cotton. The study indicated that about 20 of largest cotton organizations handle about one-third of world production. The conclusions of the study were that the world cotton industry was not highly concentrated by the standards of industrial markets and that the international cotton shipping industry was highly competitive.

The worldwide spread of Internet technology is providing an opportunity to develop online cotton marketing to improve communications and access to markets, to speed the process of marketing and to reduce overhead costs. Active efforts to design, develop and launch multiple internet-based platforms for business-to-business trade in cotton fiber and textiles started in 2000. E-commerce has the potential to increase the efficiency and speed of cotton contracting, shipping and financing, thereby facilitating the marketing process worldwide. It promises to provide wider access to price and market information to a large number of market participants. Out of all the Internet based cotton-trading platforms launched in 2000, where buyers and sellers could have an opportunity to place bids and offers, and negotiate and conclude actual delivery contracts, only The Seam based in the USA, has been active in E-commerce as of 2003 and has grown in volumes traded. DealCotton.com based in the United Kingdom and Clickforcotton.com based in India were actively developing the trading platforms initially, but had abandoned the Internet-based cotton trading platforms by 2003.

Contract for Sale of Cotton

It is customary for cotton traders to enter into agreement on the sale of cotton using telephones and verbal agreements have the same binding force as written contracts signed by parties based on the international law regulations. A verbal agreement for the sale of cotton is usually confirmed by signing a formal contract, by fax or electronic mail. It is customary to use contract forms provided by the cotton associations which are based on the association trading rules and incorporate all the major terms of sale, except for the price, origin, quality, quantity, time and place of delivery and other technical data. The American Cotton Shippers Association, Liverpool Cotton Association, Centro Algodonero Nacional (Barcelona Contract) and other associations have contract forms based on their rules that are widely used in the international cotton trade.

Contracted Quality

Cotton quality varies substantially, depending on growth, and there are several ways to stipulate quality in cotton contracts.

Sale *on description* means that cotton quality is described in the contract based on a certain standard, stipulating cotton grade, staple length and other characteristics.

Sale *on type* means that cotton quality is based on a private type of cotton, a sample of which was agreed upon as a representative quality of the lot.

Sale *on government class* means a sale made based on the government classification (example USDA) based on grade, color, staple length, micronaire and other standard measurement made by HIV. It is a practice in the USA to base sale *on Green Card*, or initial and obligatory classification of cotton performed by the USDA Classing Board when cotton is ginned. In Central Asia ginning mills issue *Certificates of Quality* with major quality characteristics of the fiber. In cases of deviation of quality from contract stipulations, value differences circulars issued by cotton associations can be used to settle differences. The associations produce the value differences estimates for use during arbitration procedures to settle quality disputes between the contract parties.

Terms of Sale

In the cotton trade, *freight* means the cost of moving the cotton from point of origin to final destination and may include the cost of moving from a warehouse or ginning mill to a shipping port, the cost of ocean transport and the cost of moving the cotton from a port of final destination to a spinning mill. A number of standard terms of sales of cotton define which party to the contract is responsible for arranging the shipment and paying the cost of transportation.

130

FOB warehouse or *ginning mill*: buyer receives cotton at a specified point and arranges for transportation from that point.

Landed mill. The seller delivers cotton to a specified mill at his own cost. This term is usually used for sales to domestic mills in cotton-producing countries.

FAS/FOB (free alongside ship/free on board) vessel: the seller pays the freight to the ocean port of loading placing cotton alongside the ship. The buyer pays the cost of ocean transportation which usually includes the cost of loading on board the ocean vessel.

C&F or CFR (cost and freight): the seller pays the ocean freight and the buyer pays the insurance.

CIF (cost, insurance and freight): the seller pays the ocean freight and the cost of marine insurance.

FOB (free on board): the seller delivers when the cotton passes the ship's railing. From that point on the cargo is at buyer's risk. Although not specifically mentioned, loading charges would be paid by the seller.

CPT (carriage paid to): the same as C&F (cost and freight), but with a delivery point located inland, not at an ocean port.

CIP (carriage and insurance paid to): same as CIF (cost, insurance and freight), but with a delivery located inland, not at an ocean port.

Shipping

Most of the standard contracts for the sale of cotton under the terms of CIF/C&F/CIPT/CIP stipulate *"freight final"* meaning that any changes in the cost of transportation between the time of entering into the contract and the time of delivery are for the sellers account. Freight rates are subject to considerable changes due to demand and supply, currency rate changes, fuel cost changes and other factors.

Most cotton shipments to standard destinations are made under *"liner terms"* clearly defined in the shipping business. Distinct from liner terms, a small percentage of cotton, usually from and/or to non-standard locations may be transported under *"charter terms."* Under charter terms, the buyer or the seller will charter all or a part of an ocean vessel negotiating each clause of the charter contract.

Until the 1970s, the only available ocean vessels were traditional *"breakbulk"* ships built to transport multi-cargo in bulk, as opposed to containers. At present, nearly all cotton shipments over water are made in containers.

Standard containers used for cotton shipping are usually 40 foot boxes containing about 19.5 metric tons of baled cotton (net weight). There are also 20 foot containers of half the capacity and used occasionally for cotton shipping. Containers have a significant advantage in efficiency over the breakbulk shipments. Cotton can be loaded into containers at a ginning mill or warehouse and transported all the way to the spinning mill overseas, reducing the cost of loading and unloading cotton during transit.

The American Cotton Shippers Association has developed a complete set of definitions and trading rules stipulating container shipments and known as *"ACSA Container Rules,"* which are in general use in the cotton trade around the world. The Liverpool Cotton Association has developed special terms associated with container operations and the *Container Trade Rules Agreement* with the American Cotton Shippers Association, which are part of the Rules and Bylaws of the Liverpool Cotton Association. Basic definitions used by the cotton shipping industry from the LCA Rule book are included here.

"Container Freight Station (CFS)" and *"container base"* means a place where the carrier or his agent loads or unloads containers under their control. Usually this would be a port or a warehouse.

"Combined transport," *"inter-modal transport"* and *"multimodal transport"* means delivering cotton from one place to another using at least two different means of transport.

"Combined transport document" means a bill of lading or other document of title produced by a shipping company, combined transport operator or agent covering cotton being moved by combined transport or multimodal transport.

"Combined transport operator" means a person or firm which produces a combined transport document.

"Container yard" (CY) means a place where containers can be parked, picked up or delivered, full or empty. A container yard or CY may also be a place where containers are loaded (or stuffed) or unloaded (or de-vanned).

"Full container load" (FCL) means an arrangement which uses all the space in a container. It is most efficient to ship in full containers since the freight is usually based on full boxes. Full container load will contain from 80 to 85 bales, meaning that this should be used as a multiple for contract quantity.

"Less than container load" (LCL) means a parcel of cotton which is too small to fill a container and which is grouped by the carrier at the container freight station with similar cargo going to the same destination.

"House to," *"container yard to,"* and *"door to"* mean loading controlled by the shipper at the place (house, CY or door) of his choice. Whoever books the freight must pay all costs beyond the point of loading and the cost of providing containers at the house, CY or door.

"Pier to," *"container freight station to,"* and *"container base to"* mean that the carrier controls the loading. The cotton must be delivered to the carrier at the pier, container freight station or container base.

"*Point of destination*" means the exact place where the cotton is delivered to the person who has ordered it, or is delivered to his agent, and where the carrier's responsibility ends.

"*Point of origin*" means the exact place where the carrier or his agent receives the cotton and where the carrier's responsibility begins.

"*Shipper's load and count*" means the shipper is responsible for the contents of the container.

"*To house,*" "*to container yard*" and "*to door*" mean delivery to the warehouse or mill yard selected by the person who booked the freight.

"*To pier,*" "*to container freight station*" and "*to container base*" mean that the carrier will unload (de-van) at his warehouse in the port of destination, in container freight station or container base.

"*Inter-modal bill of lading*" (B/L) or "*combined transport document*" mean a negotiable document issued by a water carrier after receipt of container of cotton on board a rail car or other transport equipment. A typical inter-modal B/L might cover the shipment of containerized cotton from the ginning mill or warehouse to a spinning mill overseas. It is customary to issue the ocean bill of lading or inter-modal bill of lading in three originals made out to order and blank endorsed. Since the B/L is a title document, in theory anybody holding the B/L can take the possession of cotton. The B/L might be issued "to order of…" and letting the foreign bank insert the name of the buyer after payment has been received.

On-board bill of lading means a bill which is signed by the captain of the vessel or his agent when the cotton has been loaded on the ship.

The freight rates used to be set by so-called *steamship conferences*. The conference members would agree among themselves to charge fixed freight rates from certain origins to certain destinations. There were also carriers who did not belong to a conference (non-conference carriers), usually charging lower than conference rates. The conference system gradually fell apart during the past decade. Instead, at present, it is customary for exporters and carriers to sign *volume contracts*, where shipping rates are based on volumes booked by the exporter during a certain time period with the carrier.

It is customary on many ocean routes to divide freight into several *freight components*: the base rate, bunker adjustment factor, currency adjustment factor and a terminal or origin receiving charge. The carriers may offer fixed base freight rates for the entire cotton marketing season by having the most volatile components of the freight rate stated in the form of adjustment factors.

"*Bunker adjustment factor*" (BAF), "*fuel adjustment factor*" (FAF) mean a charge added to the base freight rate to cover extraordinary increases in fuel costs which are beyond the control of the carrier.

"*Currency adjustment factor*" (CAF) means a charge, generally expressed as a percentage of base freight, that attempts to compensate for extraordinary fluctuations in currency relationships to the U.S. Dollar, which is a tariff currency.

"*Terminal receiving charge*" (TRC), terminal handling charge" (THC), "*container yard charge*" (CYC) mean a charge, added to the base freight rate by the carrier, which reflects the cost of handling cotton from place of receipt at the terminal to on board the vessel.

"*Origin receiving charge*" (ORC) means a charge, added to the base freight rate, which reflects the cost of handling cotton from place of origin to on board inter-modal conveyance.

Insurance

The party which holds title to the cotton is responsible for its insurance coverage. Normally, cotton should be covered by insurance from the time it is produced until it is consumed at the spinning mill. When cotton is shipped for export it usually is at buyer's risk once it is loaded on board the ocean vessel. In case the cotton cargo is destroyed or lost, the insurance company will pay the buyer to compensate for the loss. According to the rules of the Liverpool Cotton Association, when a buyer or seller takes out insurance on a shipment of cotton, the insurance must include:

- "*Marine cargo insurance*" and "*transit insurance*" in line with the Institute Cargo Clauses (A) or Institute Commodity Trade Clauses (A), including cover against country damage risks:
- "*Country damage*" means damage to cotton caused by moisture, dust or sand getting into the bale because it has been exposed to the weather or stored on wet or contaminated ground. Country damage occurs before shipment, as opposed to sea damage.
- "*War risk insurance*" in line with the Institute War Clauses (Cargo) or the Institute War Clauses (Commodity Trades)
- "*Strikes, riots and civil commotion's insurance*" in line with the Institute Strikes Clauses (Cargo) or Institute Strikes Clauses (Commodity Trades).

Contingency insurance means that a shipper's insurance policy should contain a contingency clause to cover the area when a title to the cotton is being transferred to the buyer. For example, damage to the cotton could occur when the cotton has been loaded on board the ocean vessel, but before the bill of lading has been issued, or before the seller has been able to collect the invoice amount.

The insurance must cover the invoice value of the shipment plus 10% (assumed profit).

Where the seller is responsible for providing marine cargo insurance or transit insurance, there must be a policy document or certificate of insurance as one of the shipping documents. When the buyer is responsible for providing marine cargo insurance or transit insurance, the buyer must insure against country damage risk, while the seller must give the buyer the necessary details for each shipment. In case of CIF sales, the insurance policy should cover the cotton for up to 30 days after the arrival at the port of destination, including delivery to the mill.

Payments and Documents

Three payment terms are commonly used in the cotton trade.

Payment against an irrevocable letter of credit (L/C) opened by the buyer in favor of the seller through the bank specified in the contract, or any first class bank in the sellers country. The seller prepares the documents in compliance with the L/C and sends them to the bank for collection. Once the payment has been made, the bank in the exporting country will forward the shipping documents to the foreign bank for transmission to the buyer.

The bulk of cotton contracts from any origin to any destination specify payment by *irrevocable L/C*, meaning that the buyer cannot withdraw the L/C once it has been opened. It is the buyer's obligation to ensure that the L/C is opened in time to be in the seller's hands by the first day of the contract shipping month. The L/C should be valid for shipment within the contract shipping month, allow partial shipments and, usually, should allow presentation of the shipping documents at the latest, 21 days after the last shipping date.

Contract terms may require a *confirmed L/C*, meaning that the bank in the seller's country that opened the L/C on behalf of the foreign bank guaranties payment even if the foreign bank defaults. The contract should specify who pays the confirmation charge.

It is the exporter's responsibility to assemble a set of *shipping documents* in strict accordance with the terms of the L/C. Usually the following documents may be required by the letter of credit:

- Seller's commercial invoice
- On board bill of lading (3 originals)
- Insurance certificate (CIF only)
- Certificate of origin
- Tag list, with the warehouse numbers for each bale
- Quality test certificate (usually HVI test values)
- Phytosanitary certificate (a government document stating that the cotton is free from diseases)
- Any other documents or declarations required by the buyer or his bank

Payment (cash) upon first presentation of the documents means that the seller sends the shipping documents to the foreign bank specified by the buyer. The buyer must remit the invoice value promptly upon receipt of the documents.

Payment (cash) upon arrival of the cotton at port of destination is similar to payment on presentation, except the payment is only due on arrival of the cotton at the port of destination.

Weighing, Sampling and Classing

Cotton can be sold under a number of terms regarding the cotton weights, such as original gin weight, inbound warehouse weight, outbound warehouse weight and net landed weight. Net landed weight is a customary term used for exports. In this case *independent controllers* must be appointed by the shipper to supervise weighing and sampling of the cotton upon arrival. The *weighing* is performed by the importer's forwarding agent upon arrival at the port of destination, and the controller supervises the procedure on behalf of the shipper.

Depending on the growth of cotton and contract terms *sampling* can be done before or after shipment. *Pre-Shipment sampling* can be done at ginning mills, warehouses or at the port of shipment by the independent controller, usually appointed by the buyer to draw an agreed percentage of samples from each lot. An average sample from a bale of cotton is between 150 to 200 grams. *Post Landed sampling* is similar procedurally to the pre-shipment sampling and is usually ordered by the seller in the event of quality disputes to have a better understanding of the nature of the claim. In case of Arbitration, the initiating party is usually responsible for drawing and forwarding the samples to the place of arbitration. The opposing party would normally appoint their own controller to supervise and seal such samples.

Samples drawn by the controller are used for quality *classing* performed often by independent controllers at the request of buyers for the purpose of confirming the quality prior to purchase or shipment in the producing country or a port of shipment. This classing may be undertaken against the national standards of the producing countries, or against the Universal Standards or against private type samples. Classing could be performed manually, with mechanical testing, or HVI equipment. Classing usually evaluates grade, color and leaf, staple length and micronaire, or a full testing by HVI.

Common Fund for Commodities Funded Projects on Cotton

This book has been published under a project sponsored by the International Cotton Advisory Committee (ICAC) and funded by the Common Fund for Commodities (CFC). The CFC is an inter-governmental financial institution established within the framework of the United Nations and based in Amsterdam, The Netherlands. It is mandated to enhance socio-economic development by concentrating on general, commodity-wide aspects and problems of commodity development. ICAC is the recognized International Commodity Body on cotton and sponsors projects for financing by the Common Fund. Since 1991, when the Fund approved the first project for funding, fifteen projects have been funded in cotton until 2003 and others are under consideration for final support (as at end 2003). These projects are:

1. *Cotton Production Prospects for the Next Decade (CFC/ICAC 01)*
   Duration: November 1992 to March 1995
   Countries: Brazil, China (Mainland), Egypt, India, Mali, Mexico, Pakistan, Tanzania and Uzbekistan
   Project Executing Agency: The World Bank

2. *Integrated Pest Management for Cotton (CFC/ICAC 03)*
   Duration: September 13, 1994 to September 30, 1999
   Countries: Egypt, Ethiopia, Israel and Zimbabwe
   Project Executing Agency: The Israeli Cotton Production and Marketing Board Ltd

3. *Integrated Pest Management of the Cotton Boll Weevil in Argentina, Brazil and Paraguay (CFC/ICAC 04)*
   Duration: June 30, 1995 to June 30, 2001
   Countries: Argentina, Brazil and Paraguay
   Project Executing Agency: National Service for Phytosanitary and Agro Food Quality, Argentina

4. *Genome Characterization of Whitefly-Transmitted Geminivirus of Cotton and Development of Virus-Resistant Plants through Genetic Engineering and Conventional Breeding (CFC/ICAC 07)*
   Duration: January 1, 1996 to December 31, 2001
   Countries: Pakistan, UK and USA
   Project Executing Agency: National Institute for Biotechnology and Genetic Engineering, Pakistan

5. *Improvement of the Marketability of Cotton Produced in the Zones Affected by Stickiness (CFC/ICAC 11)*
   Duration: January 1, 1997 to April 30, 2001
   Countries: France and Sudan
   Project Executing Agency: The Sudan Cotton Company Ltd.

6. *Improvement of Cotton Marketing and Trade Systems in Eastern and Southern Africa (CFC/ICAC 12)*
   Duration: October 1, 2000 to September 30, 2003
   Countries: Uganda and Tanzania
   Project Executing Agency: United Nations Office of Project Services

7. *Sustainable Control of the Cotton Bollworm Helicoverpa armigera in Small-scale Cotton Production Systems (CFC/ICAC 14)*
   Duration: October 1, 2000 to September 30, 2004
   Countries: China (Mainland), India, Pakistan and UK
   Project Executing Agency: Natural Resources International Ltd. UK

8. *Resistance Management of Helicoverpa armigera to Pyrethroids in West Africa (CFC/ICAC/16)*
   Duration: Fast track project (2000-2001)
   Countries: Benin, Burkina Faso, Cote d'Ivoire, Mali, Senegal, Togo, Nigeria and Guinee
   Project Executing Agency: Institut de l'Environnement et de Recherches Agricoles, Burkina Faso

9. *Pilot Project on Price Risk Management for Cotton Farmers (CFC/ICAC 17)*
   Duration: Three year
   Countries: Tanzania, Uganda and Zimbabwe

10. *Pilot Project on Price Risk Management for Cotton Farmers (CFC/ICAC 19FT)*
    Duration: January 2002-March 2003
    Countries: Tanzania, Uganda and Zimbabwe

11. *Assessment of the Impact and Main Dynamics of Cotton Diseases Affecting in Particular Small Scale Production Systems in Eastern and Southern Africa (CFC/ICAC 21FT)*
    Duration: May 1, 2002 to October 30, 2002
    Countries: Ethiopia, South Africa, Sudan, Tanzania, Uganda and Zimbabwe
    Project Executing Agency: Secretariat of the Southern and Eastern African
    Cotton Forum (SEACF)

12. *Cotton Facts (CFC/ICAC 23FT)*
    Duration: May 1, 2002 to April 30, 2003
    Project Executing Agency: International Cotton Advisory Committee

13. *Improvement of the Sustainability of Cotton Production in West Africa (CFC/ICAC 25FT)*
    Duration: June 1, 2003 to December 31, 2003
    Countries: Benin, Burkina Faso, Cameroon, Chad, Ivory Coast, Mali and Togo
    Project Executing Agency: United Nations Conference on Trade and
    Development (UNCTAD), Geneva, Switzerland

14. *Utilization of Cotton By-produce for Value-added Products (CFC/ICAC 20)*
    Duration: Four years
    Countries: India
    Project Executing Agency: Central Institute for Research on Cotton
    Technology, India

15. *Utilization of Cotton By-produce for Value-added Products (CFC/ICAC 27FT)*
    Duration: May/June 2003
    Countries: India
    Project Executing Agency: Central Institute for Research on Cotton
    Technology, India

COMMON FUND FOR COMMODITIES

The Common Fund for Commodities (CFC) is an autonomous intergovernmental financial institution established within the framework of the United Nations. The Agreement Establishing the Common Fund for Commodities was negotiated in the United Nations Conference on Trade and Development (UNCTAD) from 1976 to 1980 and became effective in 1989. The Common Fund for Commodities forms a partnership of 106 Member States plus the European Community (EC), the African Union (AU) and the Common Market for Eastern and Southern Africa (COMESA) as institutional members. The Common Fund's mandate is to enhance the socio-economic development of commodity producers and contribute to the development of society as a whole. The Common Fund operates under the novel approach of commodity focus instead of the traditional country focus. Commodity focus entails concentrating on the general problems of commodities of interest to developing countries. The activities of the Fund mainly comprise of commodity development measures aimed at improving the structural conditions in markets and enhancing the long-term competitiveness and prospects of particular commodities. The Fund assists developing countries, in particular least developed countries (LDCs) and countries with economies in transition, to function effectively in a liberalized global economy.

Common Fund for Commodities
P. O. Box 74656
1070BR Amsterdam
The Netherlands
Telephone: 31-20-5754949
Fax: 31-20-6760231
http://www.common-fund.org/

INTERNATIONAL COTTON ADVISORY COMMITTEE _____

The International Cotton Advisory Committee (ICAC) is an association of governments of cotton producing, consuming and trading countries. The Committee was formed in 1939, and the secretariat was established in 1946. The mission of the ICAC is to assist governments in fostering a healthy world cotton economy, and the work of the Committee is important to government officials and all segments of the private sector involved with cotton. The Committee achieves its mission by providing transparency to the world cotton market, by serving as a clearinghouse for technical information on cotton production and by serving as a forum for discussion of cotton issues of international significance. The role of the ICAC is to raise awareness of emerging issues, provide information relevant to the solving of problems and to foster cooperation in the achievement of common objectives. The ICAC is the premier source of international data on the world cotton industry. The Secretariat forecasts cotton supply, use and prices, estimates cotton supply by type, and tracks exports by destination and imports by origin. The Secretariat measures and forecasts cotton consumption and cotton's share of fiber demand in the world and by region and is the primary source in the world for statistics on fiber demand. The ICAC is the International Commodity Body for cotton before the Common Fund for Commodities.

International Cotton Advisory Committee
1629 K Street, N.W., Suite 702
Washington DC, 20006
USA
Telephone: 202-463-6660
Fax: 202-463-6950
http://www.icac.org/

Conversion Factors

WEIGHT

	Kilograms	Pounds	480 lb bales
Metric ton	1000	2204.6	4.593
Pound	0.4593	1	480
Kilogram	1	2.2046	0.004593
Arroba (Brazil)	15	33.069	0.068895
Candy (India)	355.62	784	1.6333
Cantar, metric (Egypt)	50	110.23	0.2296
Cantar (Sudan)	44.93	99.04	0.20635
Centner (USSR)	100	220.46	0.4593
Dan (China)	50	110.23	0.2296
Quintal (Argentina)	45.95	101.3	0.211
Quintal (India)	100	220.46	0.4593
Quintal (Mexico)	46.026	101.47	0.2113
Quintal (Peru, Spain)	46	102.43	0.2113
Long ton	1016	2240	4.666
Maund (Pakistan)	37.3242	82.286	0.1714
Picul (China)	50	110.23	0.2296

BALES

Australia	227	500	1.04167
Colombia	233	514	1.0702
Egypt	327	720	1.5
India/Pakistan	170	375	0.7808
Mexico	230	507	1.05625
Nigeria	185	408	0.85
South Africa	200	441	0.9186
Sudan	191	420	0.875
Tanzania/Uganda	181	400	0.83333
USA	225	496	1.033

AREA

	Acres	Hectares
Acre	1	0.40469
Dunams	0.24710	0.1
Feddan	1.038	0.42
Hectare	2.47103	1
Manzana	1.72	0.696
Mu (China, 1/15 hectare)	0.16473	0.06667
Stremma (Greece, 1/10 hectare)	0.24710	0.1

TO CONVERT FROM:	To:	Multiply by:
LENGTH		
Centimeter	Inch	0.3937
Centimeter	Millimeter	10
Inch	Centimeter	2.54
Inch	Millimeter	25.4
Foot	Centimeter	30.48
Kilometer	Mile	0.6215
Meter	Inch	39.37
Meter	Yard	1.09361
Meter	Foot	3.2808
Mile	Kilometer	1.60934
Millimeter	32nd of an inch	1.25984
Yard	Meter	0.9144
32nd of an inch	Millimeter	0.79375
YIELDS		
Kilograms per hectare	Pound per acre	0.89219
Cantar per feddan	Pound per acre	106.1895
Cantar per feddan	Kilogram per hectare	94.74
Quintal per manzana	Pounds per acre	58.9709
Quintal per manzana	Kilogram per hectare	0.0169575
Pounds per acre	Kilogram per hectare	1.12084
PRICE		
Cents per pound	Dollars per ton	22.046
Dollars per ton	Cents per pound	0.04536
OTHER		
Kilogram per sq. meter	Pound per sq. yard	1.84336
Pound per sq. yard	Kilogram per sq. meter	0.54249
Square meter	Square yard	1.19603
Square yard	Square meter	0.8361
Gram	Ounce	0.0353

Cotton Lint Supply and Use in World Total

Years Beginning August 1	Area 000 Ha	Yield Kgs/ Ha	Prod	Beg Stks	Impts	Cons	Expts	End Stks	S/U*
					000 Metric Tons				
1950/51	28,537	234	6,674	3,708	2,724	7,638	2,673	2,678	0.35
1951/52	36,040	234	8,426	2,678	2,663	7,657	2,712	3,417	0.45
1952/53	35,448	246	8,730	3,417	2,613	8,044	2,623	4,070	0.51
1953/54	33,422	272	9,076	4,070	2,878	8,443	2,950	4,626	0.55
1954/55	33,445	267	8,939	4,626	2,757	8,678	2,716	4,862	0.56
1955/56	34,078	279	9,511	4,862	2,882	8,972	2,853	5,349	0.60
1956/57	33,417	277	9,240	5,349	3,409	9,352	3,508	5,113	0.55
1957/58	32,032	283	9,051	5,113	3,066	9,317	3,114	4,803	0.52
1958/59	31,657	308	9,758	4,810	3,043	9,942	2,960	4,609	0.46
1959/60	32,326	318	10,280	4,609	3,789	10,529	3,818	4,407	0.42
1960/61	32,445	313	10,154	4,407	3,811	10,231	3,716	4,643	0.45
1961/62	33,059	296	9,796	4,643	3,463	9,982	3,398	4,531	0.45
1962/63	32,638	319	10,424	4,531	3,638	9,822	3,467	5,268	0.54
1963/64	32,974	330	10,867	5,268	3,887	10,362	3,935	5,852	0.56
1964/65	33,367	345	11,504	5,852	3,811	11,145	3,721	6,317	0.57
1965/66	33,133	359	11,895	6,317	3,809	11,407	3,712	6,839	0.60
1966/67	30,915	351	10,836	6,839	3,934	11,598	3,974	6,079	0.52
1967/68	30,670	351	10,780	6,079	3,828	11,731	3,805	5,044	0.43
1968/69	31,692	374	11,856	5,044	3,696	11,753	3,640	5,231	0.45
1969/70	32,657	348	11,379	5,231	3,932	11,983	3,880	4,700	0.39
1970/71	31,778	369	11,740	4,656	4,086	12,173	3,875	4,605	0.38
1971/72	33,024	392	12,938	4,681	4,031	12,721	4,111	4,799	0.38
1972/73	33,818	402	13,595	4,851	4,528	13,034	4,640	5,358	0.41
1973/74	32,558	418	13,615	5,434	4,408	13,469	4,294	5,727	0.43
1974/75	33,285	418	13,924	5,727	3,734	12,641	3,814	7,373	0.58
1975/76	30,003	390	11,705	7,352	4,188	13,336	4,183	5,770	0.43
1976/77	31,512	393	12,384	5,770	3,951	13,122	3,806	5,232	0.40
1977/78	34,965	396	13,859	5,232	4,250	13,133	4,239	5,963	0.45

* Ending stocks divided by consumption plus exports.

Years Beginning August 1	Area 000 Ha	Yield Kgs/ Ha	Prod	Beg Stks	Impts	Cons	Expts	End Stks	S/U*
					000 Metric Tons				
1978/79	33,999	380	12,932	5,963	4,320	13,703	4,346	5,257	0.38
1979/80	33,099	425	14,083	5,256	5,093	14,127	5,073	5,258	0.37
1980/81	33,667	411	13,831	5,152	4,555	14,215	4,414	4,994	0.35
1981/82	33,948	442	14,991	4,994	4,405	14,147	4,373	5,852	0.41
1982/83	32,569	445	14,479	5,852	4,350	14,452	4,261	5,926	0.41
1983/84	32,137	451	14,499	5,926	4,617	14,655	4,309	6,121	0.42
1984/85	35,217	547	19,247	6,121	4,602	15,108	4,520	10,247	0.68
1985/86	32,792	532	17,461	10,247	4,763	16,589	4,479	11,366	0.69
1986/87	29,503	518	15,269	11,366	5,516	18,198	5,755	8,251	0.45
1987/88	31,238	564	17,609	8,251	5,094	18,117	5,121	7,668	0.42
1988/89	33,522	546	18,301	7,668	5,654	18,470	5,726	7,312	0.40
1989/90	31,640	549	17,365	7,312	5,431	18,675	5,293	6,146	0.33
1990/91	33,049	574	18,978	6,146	5,220	18,574	5,073	6,709	0.36
1991/92	34,710	596	20,677	6,709	6,497	18,637	6,091	9,311	0.50
1992/93	32,248	556	17,941	9,312	5,690	18,635	5,525	8,690	0.47
1993/94	30,435	554	16,861	8,690	5,766	18,496	5,911	7,024	0.38
1994/95	32,112	584	18,762	7,024	6,458	18,378	6,312	7,456	0.41
1995/96	36,066	564	20,330	7,456	5,808	18,456	5,999	8,975	0.49
1996/97	34,195	573	19,584	8,975	6,140	19,094	6,042	9,644	0.51
1997/98	33,850	593	20,074	9,644	5,744	19,033	5,968	10,421	0.55
1998/99	32,893	569	18,712	10,423	5,414	18,472	5,505	10,699	0.58
1999/00	31,953	597	19,090	10,699	6,055	19,634	6,107	10,107	0.51
2000/01	31,930	609	19,461	10,107	5,748	19,874	5,855	9,707	0.49
2001/02	33,497	642	21,514	9,707	6,148	20,185	6,471	10,613	0.53
2002/03, est.	30,131	639	19,263	10,613	6,405	20,871	6,647	8,735	0.42
2003/04, for.	32,635	623	20,316	8,735	6,722	20,828	6,722	8,223	0.39
2004/05, for.	34,135	639	21,816	8,223	6,572	20,815	6,572	9,223	0.44

* Ending stocks divided by consumption plus exports.

References

Afzal, Muhammad. 1983. *Cotton Plant in Pakistan*, Second Edition, Ismail Aiwan-i-Science, Sharrah-i-Roomi, Lahore, Pakistan.

Afzal, Muhammad. 1986. *Narratio Botanica Concerning the Yield of Crops*, Shah Enterprises. P. O. Box 6582, 2nd Floor, Ebrahim Building, 20 West Wharf Road, Karachi, Pakistan.

Bafes, John. 2002. *The Rise and Fall of Cotton Exchanges*, March 2002, Washington, D.C., USA.

Basra, Amarjit S. 1999. *Cotton Fibers – Developmental Biology, Quality Improvement, and Textile Processing*, Food Products Press®, an imprint of The Haworth Press, Inc., 10 Alice Street, Binghamton, NY 13904-1580, USA.

Basu, A. K., Iyer, K. R. Krishna, Narayanan, S. S. and Rajendran, T. P. 1999. *Handbook of Cotton in India*, Indian Society for Cotton Improvement, c/o Central Institute for Research on Cotton Technology, Adenwala Road, Matuwa, Mumbai 400019, India.

Behery, H. M. 1993. *Short Fiber Content and Uniformity Index in Cotton*, ICAC Review Article on Cotton Production Research No. 4, CAB International, Marketing and Distribution Services, Wallingford, Oxon OX10 8DE, UK.

Castle, Steve J., Prabhaker, Nilima and Henneberry, Thomas J. 1999. *Insecticide Resistance and its Management in Cotton Insects*, ICAC Review Article on Cotton Production Research No. 5. Technical Information Section, International Cotton Advisory Committee, 1629 K Street, Suite 702, Washington, D.C. 20006, USA.

Cauquil, Jean. 1988. *Cotton Pests and Diseases in Africa South of Sahara*, Centre de Coopération Internationale en Recherche Agronomique pour le Développement (CIRAD), 34398 Montpellier Cedex 5, France.

Chaudhry, Rafiq M. 1991-2003. THE ICAC RECORDER, Technical Information Section, International Cotton Advisory Committee, 1629 K Street, Suite 702, Washington, D.C. 20006, USA.

Constable, G. A. and Forrester, N. W. 1995. *Challenging the Future*, Proceedings of the World Cotton Research Conference-1 held in Brisbane, Australia from February 14-17, 1994, International Cotton Advisory Committee, 1629 K Street, Suite 702, Washington, D.C. 20006, USA.

Cotton Outlook, 2003, *The Cotlook Indices*, A Brief Description, Liverpool, UK.

Davis, Ted. 2003. *The New York Cotton Exchange, The Long and the Short View of Cotton Futures*, New York Board of Trade, New York, USA.

Frisbie, Raymond E., El-Zik, Kamal M. and Wilson, L. Ted. 1989. *Integrated Pest Management Systems and Cotton Production*, John Wiley & Sons, New York, USA.

Gillham, Fred M. 2000. *New Frontiers in Cotton Research*, Proceedings of the World Cotton Research Conference-2 held in Athens, Greece from September 6-12, 1998, International Cotton Advisory Committee, 1629 K Street, Suite 702, Washington, D.C. 20006, USA.

Green, M. B. and Lyon, D. J. de B. 1989. *Pest Management in Cotton*, Ellis Horwood Limited, Market Cross House, Cooper Street, Chichester, West Sussex, PO 19 1EB, England.

Hillock, R. J. 1992. *Cotton Diseases*, C.A.B. International, Wallingford, Oxon OX10 8DE, UK.

Hirschfeld, Peter E. 2003. *Freight and Insurance, Payments and Documents*, Dallas, USA.

Ingram, W. R. 1981. *Pests of West Indian Sea Island Cotton*, Centre for Overseas Pest Research, College House, Wright Lane, London W8 5SJ, UK.

King, Edgar G., Phillips, Jacob R. and Coleman, Randy J. 1996. Cotton Insects and Mites: Characterization and Management, The Cotton Foundation, 1918 North Parkway, Memphis, TN 38112, USA.

Knutson, Allen and Ruberson, John. 1996. *Recognizing the Good Bugs in Cotton: Field Guide to Predators, Parasites and Pathogens Attacking Insect and Mite Pests of Cotton*, Publication and Supply Distribution, Texas Agricultural Extension Service, P. O. Box 1209, Bryan, TX 77806-1209, USA.

Kohel, R. J. and Lewis, C. F. 1984. *Cotton*, Number 2 in the series Agronomy, American Society of Agronomy, Inc., Crop Science Society of America, Inc., Soil Science Society of America, Inc., Publishers, 677 South Segoe Road, Madison 53711, Wisconsin, USA.

Liverpool Cotton Association. 2003. *Bylaws and Rules of the Liverpool Cotton Association*, Liverpool, UK.

Mathews, G. A. and Tunstall, J. P. 1994. *Insect Pests of Cotton*, CAB International, Wallingford, Oxon OX10 8DE, UK.

Mauney, Jack R. and Stewart, James McD. 1986. *Cotton Physiology*, The Cotton Foundation, 1918 North Parkway, Memphis, TN 38112, USA.

Munro, John M. 1987. *Cotton*, Longman Scientific & Technical, Longman Group UK Limited, Longman House, Brunt Mill, Harlow, Essex CM20 2JE, England.

National Cottonseed Products Association, Inc. 1990. *Cottonseed and its Products*, National Cottonseed Products Association, Inc. P. 0. Box 172267, Memphis, TN 38187-2267, USA.

New York Board of Trade. 2003. *Understanding Futures and Options*, other pamphlets and materials of the New York Board of Trade, New York, USA.

Pakistan Central Cotton Committee. 1988. *Cotton in Pakistan*, Pakistan Central Cotton Committee, Ministry of Food and Agriculture, Government of Pakistan, Moulvi Tamizuddin Khan Road, Karachi, Pakistan (Urdu).

Puri, S. N., Sharma, O. P. Murthy, K. S. and Raj, Sheo. 1998. *Hand Book on Diagnosis and Integrated Management of Cotton Pests*, National Centre for Integrated Pest Management, Pus Campus, New Delhi-110012, India.

Saha, Sukumar and Jenkins, Johnie N. 2001. *Genetic Improvement of Cotton: Emerging Technologies*, Science Publishers, Inc., Enfield, New Hampshire, USA.

Scott, Nigel. 2003. SWAPS, Rabobank International, Utrecht.

Seagull, Robert and Alspaugh, Pam. 2001. *Cotton Fiber Development and Processing – An Illustrated Overview*, International Textile Center, Texas Tech University, Lubbock, Texas, USA.

Semongulian, N G. 1991. *Genetics of Qualitative Characteristics of Cotton*, FAN Publishing House, Tashkent, Uzbekistan (Russian).

Stewart, J. McD. 1991. *Biotechnology of Cotton*, ICAC Review Article on Cotton Production Research No. 3, CAB International, Marketing and Distribution Services, Wallingford, Oxon OX10 8DE, UK.

Technical Information Section. 1991-2002. Papers presented at the Technical Seminars held on various topics at the ICAC Plenary Meetings from 1991-2002, International Cotton Advisory Committee, 1629 K Street, Suite 702, Washington, D.C. 20006, USA.

Wakefield, Peter. 2003. *Weight and Quality Control*, Wakefield Inspection Services, Liverpool, UK.

Index

London Commodity Exchange, *112*

Magnesium, *42*

Mass selection, *20*

Maturity, *6, 11, 36, 40-41, 62, 65, 68, 88*

Mean length, *87-88*

Mechanical ricker, *92-93*

Mercerization, *86, 99*

Merchants, *101, 107, 110, 113, 119-122, 124,-125, 129*

Micronaire, *34, 88, 95, 114, 130, 136*

Modal length, *87*

Module system, *91*

Monogenic, *19*

Monopodial branches, *3-4, 38*

Motes, *6, 89, 92, 94*

Multi-adversity resistance, *83*

Mutagen, *18*

Mutation breeding, *18*

Natural enemies, *63, 73-76, 79-80, 83*

    Parasites, *61, 73, 75-76, 79, 83-84, 143*
      Encarsis spp., *77*
      Eretmocerus spp., *77*
      Trichogramma spp., *63, 76*
    Pathogens, *29, 43, 45-46, 51-52, 55, 73, 77, 79, 83, 146*
      Bacillus thuringiensis, 26, 31, 77
      Beauveria bassiana, 77
      Erynia spp., *77*
      Nuclear polyhedrosis viruses, 77
      Nomouraea rileyi, 77
    Predators, *59, 71, 74-75, 79, 83, 148*
      Catolaccus grandis, 74
      Chrysopa spp., *73*
      Geocoris spp., *73*
      Hippodamia spp., *73*
      Nabis spp., *74*
      Orius spp., *74*
      Pseudatomoscelis seriatus, 74
      Solenopsis spp., *74*
      Zelus and Sinea spp., *74*

Spiders, *75-76*
    Acanthepeira stellata, 75
    Aysha gracilis, 75
    Chiracanthium inclusum, 75
    Misumenops celer, 75
    Oxyopes salticus, 76
    Phidippus audax, 76

Natural out crossing, *17-19, 32-33*

Natural selection, *17, 20*

Nep, *93*

New Orleans Cotton Exchange (NOCE), *110-112*

New world cotton, *2*

New York Board of Trade (NYBOT), *105-107, 110, 113, 148-149*

New York Cotton Exchange (NYCE), *110-114, 148*

Nitrogen, *3, 12, 37, 39-42, 48, 52*

Nodes, *3, 36-37, 59, 61*

Noil, *97*

Nonwoven fabric, *99*

Nutrition, *11-12, 40*

Nymph, *58, 77, 83*

Oil, *xiii, 6-9, 33, 60*

Okra, *11, 58, 60, 79*

Old world cotton, *2*

Organic, *12, 15-16, 30, 40, 42, 128*

Osaka Sampin Exchange, *111*

Oviposition, *58-59, 83*

Parthenogenesis, *57, 84*

Payments and documents, *135, 145*
    Confirmed letter of credit (L/C), *135*
    Irrevocable letter of credit (L/C), *135*
    Payment (cash) upon arrival of the cotton at port of destination, *136*
    Payment (cash) upon first presentation of the documents, *136*
    Shipping documents, *135-136*

Pedigree selection/Progeny row selection, *19*

Phenotype, *17, 21*

Ring spinning, *95, 97*

Roller gin, *92*

Root, *5-7, 14, 27, 39, 41, 45-47, 51-52*

Rosette flower, *84*

Rotor spinning, *95-96*

Russian Cotton Association (RCA), *125*

Sale of cotton, *104, 119, 130-131*
    On description, *130*
    On government class, *130*
    On green card, *130*
    On type, *130*

Salinity, *12, 37*

Sampling, *136*

Sanforization, *99*

Saw gin, *92*

Sea Island cotton, *2, 145*

Secondary outbreak, *84*

Seed, *xiii, 5-10, 12, 14, 19-22, 30-33, 37, 39, 41, 45-46, 48-49, 51, 55, 60, 79, 85, 89, 91-93, 129*

Seed coat fragment, *89*

Selection coefficient, *20*

Selfing, *20*

Shanghai Cotton Exchange, *111*

Shedding. *12, 37-38, 40, 48, 53, 59, 61, 66, 91*

Shipping, *xiii, 119, 123, 129-136*
    ACSA Container Rules, *132*
    Breakbulk, *131-132*
    Bunker adjustment factor (BAF), *133*
    Charter terms, *131*
    Combined transport, *132-133*
    Combined transport document, *132-133*
    Combined transport operator, *132*
    Container base, *132-133*
    Container base to, *132*
    Container freight station (CFS), *132-133*
    Container freight station to, *132*
    Container Trade Rules Agreement, *132*
    Container yard (CY), *132-134*
    Container yard charge (CYC)
    Container yard to, *132*
    Currency adjustment factor (CAF), *133-134*
    Door to, *132*
    Freight components, *133*
    Freight final, *131*
    Fuel adjustment factor (FAF), *133*
    Full container load (FCL), *132*
    House to, *132*
    Inter-modal bill of lading (B/L), *132*
    Inter-modal transport, *132*
    Less than container load (LCL), *132*
    Liner terms, *131*
    Multimodal transport, *132*
    On-board bill of lading (B/L), *133*
    Origin receiving charge (ORC), *133-134*
    Pier to, *132*
    Point of destination, *133*
    Point of origin, *130, 133*
    Shipper's load and count, *133*
    Standard containers, *131*
    Steamship conferences, *133*
    Terminal handling charge (THC), *134*
    Terminal receiving charge (TRC), *134*
    To container base, *133*
    To container freight station, *133*
    To container yard, *133*
    To door, *133*
    To house, *133*
    To pier, *133*
    Volume contracts, *133*

Short fiber content, *89, 91, 93-94, 100, 147*

Single cross, *19*

Slubbing and roving frames, *97*

Solarization, *52, 84*

Southern Mill Rules, *107-108, 118*

Spindle harvester, *90*

Spinning, *88, 92, 94-100, 102, 119-121, 123, 130, 132-134*

Square, *36-39, 58-60, 63-64, 70-72, 74-75, 99, 139*

Staple length, *87, 91, 93, 114, 130, 136*